Infectious Diseases and Microbiology

nova
Medicine & Health
New York

Infectious Diseases and Microbiology

Psychological Consequences of the COVID-19 Pandemic on Children, Teenagers and Adults
Jean-Pascal Assailly, PhD (Editor)
2022. ISBN: 978-1-68507-963-5 (Softcover)
2022. ISBN: 979-8-88697-072-2 (eBook)

COVID-19 Contracting: Opportunities, Actions and Reimbursements
Eric Neudorf (Author)
2022. ISBN: 978-1-68507-710-5 (Hardcover)
2022. ISBN: 978-1-68507-733-4 (eBook)

The Aviation Industry and COVID-19
Stefan Schneider (Author)
2022. ISBN: 978-1-68507-712-9 (Hardcover)
2022. ISBN: 978-1-68507-729-7 (eBook)

Diseases Transmitted by Ticks
Hacer İşler (Editor), Metin Özdemir, MD (Editor)
2021. ISBN: 978-1-68507-359-6 (Hardcover)
2021. ISBN: 978-1-68507-383-1 (eBook)

Impacts and Implications of COVID-19: An Analytical and Empirical Study
Anand Sharma (Editor)
Prateek Agrawal (Editor)
Vishu Madaan (Editor)
Anuj Agarwal (Editor)
2021. ISBN: 978-1-53619-670-2 (Hardcover)
2021. ISBN: 978-1-68507-081-6 (eBook)

More information about this series can be found at
https://novapublishers.com/product-category/series/infectious-diseases-and-microbiology/

Digvijay Pandey, Binay Kumar Pandey,
Nidhi Sindhwani, Rohit Anand,
Vinay Kumar Nassa and
Pankaj Dadheech
Editors

An Interdisciplinary Approach in the Post-COVID-19 Pandemic Era

nova
Medicine & Health
New York

https://doi.org/10.52305/KCJU7458

Library of Congress Cataloging-in-Publication Data

ISBN: 979-8-88697-052-4

Published by Nova Science Publishers, Inc. † New York

Contents

Preface

The coronavirus family is a broad family of viruses that cause sickness in both animals and humans. Coronavirus has been linked to a variety of human respiratory infections, from the common cold to more serious illnesses, including Middle East respiratory syndrome (MERS) and severe acute respiratory syndrome (SARS). Coronavirus illness (COVID-19) was caused by a newly discovered coronavirus that swept the globe, killing hundreds of thousands of people, shutting down businesses, closing borders, and spreading misery on an unparalleled scale. In addition to its influence on public health, COVID-19 has had a substantial economic impact. As a result, doing academic research on numerous elements of this highly infectious disease is urgently needed in order to identify effective measures of disease containment and treatment, both now and in the future.

The major goal of this book is to raise awareness of COVID-19 and its influence on society, as well as to identify research fields linked to it. It might contribute to a better understanding of the sickness and characterise the psychological effects of the pandemic, as well as how these might alter as the epidemic spreads. COVID-19's negative effects will be examined in this book, along with their benefits, drawbacks, and future directions.

This book will also examine the social, cultural, and economic effects of the pandemic on these countries, as well as how the disease is affecting the public's mental health. It will also look at how government response policies have influenced the pandemic's spread, as well as the role of social media in disseminating vital information to help contain the virus's global spread.

The economic influence of COVID-19 on ten other significant industries is also examined in this book, including financial institutions, agriculture and supply chains, tourism and hospitality, FMCG, pharmaceuticals, education, railways, Delhi train services, power, and telecom. There are also some suggested techniques for improving the epidemic and dealing with thc adverse scenarios created by COVID-19. Finally, we'll go through several categorization algorithms using machine learning techniques such as Support Vector Machine (SVM), Random Forest (RF), and Naive Bayes (NB).

The investigation demonstrates that the support vector machine (SVM) approach is optimal for accuracy, while the random forest (RF) method is best for sensitivity, to provide optimum prediction results to evaluate the influence of COVID-19 on many vital industries. This book was co-authored by a group of researchers and academicians around the globe who offer a variety of perspectives and answers to some of the most pressing current issues.

Chapter 1 is entitled "Assessing the Impacts, Trends and Responses to COVID-19 in Developing Countries using Machine Learning Techniques". In this chapter, the writers examine the social, cultural, and economic effects of the COVID-19 pandemic on countries, as well as how the sickness affects people's mental health. Finally, they investigate how government response strategies have influenced the spread of the pandemic, as well as the role of social media in disseminating critical information to aid in the containment of the virus's worldwide spread.

Chapter 2 is entitled "Students' Behavior in Facing Cultural Arts to Support Child Friendly School (CFS): Preserving Local Culture at Schools during COVID-19 Crisis in Gayo, Aceh". During the COVID-19 crisis, the writers highlight the students' involvement in protecting local cultural arts at school to organise, administer, and assess. The research is descriptive and qualitative in nature. The study's participants are 78 primary school pupils from Gayo, Aceh, Indonesia. They come from a number of different primary schools. Informants and documents are the data sources. Elementary students are the informants. The paper is a teacher's lesson plan. Questionnaires and document analysis are the methods used to acquire data.

Chapter 3 is entitled "Impact of COVID -19 on Various Sectors". The influence of COVID-19 on financial institutions, agriculture and supply chains, tourism and hospitality, FMCG, pharmaceuticals, education, railways, Delhi rail services, power, and telecom was examined in this chapter.

Chapter 4 is entitled "Consequence of COVID-19 Pandemic on Web-based Educational Process: Benefits, Problems and Suggestions". This is a condensed chapter in which the authors address the different challenges that teaching institutions confront throughout the online learning and evaluation process in the context of the COVID-19 epidemic. It also offers some suggestions for how to improve the web-based learning and assessment system, as well as how to cope with troublesome circumstances. This report also discusses the advantages of e-learning and how it may help reshape the future of education.

Chapter 5 is entitled “Evaluation of the Sars-Cov-2 Omicron Variant's Features, Prospective Consequences, and Upcoming Prospects”. In this chapter, the authors describe how COVID-19 caused significant damage to worldwide wellbeing. In South Africa, a new level severe respiratory syndrome coronavirus 2 (SARS-Cov-2) variant, B.1.1.529, was found, resulting in a substantial increase in COVID-19 patients.

Chapter 6 is entitled “Impact on Various Sectors During COVID-19 Crisis”. The writers discuss the topic of COVID-19's fast-increasing influence on the global economy, which results in job losses, economic disruptions, and makes us reflect on one of the worst periods in human history. Almost every industry is seeing a dramatic drop in sales, and the pandemic's impact is so severe that much worse times ahead are anticipated. The current chapter has focused on the many industries that have been most affected by COVID-19, which has resulted in job losses and a worldwide economic downturn.

Chapter is 7 entitled “COVID-19 and Post-COVID Effect in India: A Critical Analysis”. The author discusses the post-COVID effect, as well as the state of the government and personnel dealing with the crisis. The current chapter examines the current state of information about COVID-19, as well as numerous post-COVID-19 symptoms recorded across the world, particularly in India, in terms of professional position, financial status, mental status, and the government's and people's responses to the pandemic.

Chapter 8 is entitled “Detection and Analysis of Mental Stress in Online Learning using IoT Devices”. This is a summary chapter in which a model is constructed and developed to identify stress levels using multiple sensors such as heart rate, blood pressure, body temperature, and CO2 gas concentration. The levels of stress are then estimated based on the measurements of these sensors, and the data is relayed via IoT.

Chapter 9 is entitled “COVID-19: The Enemy Threatening Mental Health Among Medical Warriors”. The authors discussed the constraints imposed by the nature of labour in health-care institutions during the pandemic, as well as the major psychiatric disorders created by those burdens, citing some of the cross-sectional studies conducted throughout the world.

Chapter 10 is entitled “Environmental Behavior Analysis and Monitoring with IoT for COVID-19”. This is a summary chapter in which a model is constructed and developed to identify stress levels using multiple sensors such as heart rate, blood pressure, body temperature, and CO2 gas concentration. The levels of stress are then estimated based on the measurements of these sensors, and the data is relayed via IoT. The authors reviewed the most common uses of sensors for monitoring COVID-19 in this chapter, including proximity sensors, temperature sensors, smoke sensors, and so on. With an ever-growing population, increased industrial activity, and rising energy consumption, it's critical to improve environmental monitoring accuracy and efficiency by using cutting-edge, automated monitoring apps and devices. It will be a great tool for climate change monitoring. As a consequence, people, energy, products, and information will flow more efficiently. It will also continue to evolve as scientists and leaders reach fresh agreements on the activities that must be undertaken to protect us from increasing temperatures and dangerous climatic conditions.

Chapter 11 is entitled "A Study on Mental Stress of Online Learning during COVID-19". According to the researchers, the globe is on the verge of experiencing the worst health disaster in modern history. All nations have been brought to their knees by the coronavirus (COVID-19), which has shut down many sectors and businesses, including education. Teachers, students, and other education stakeholders are debating what would happen to education if a majority of governments throughout the world were given the ability to divide society as part of a "new norm." As a result, educational institutions are building an online platform for teaching and guaranteeing that students continue to pursue their studies across a variety of social barriers, but this has an impact on students' mental stress.

Chapter 12 is entitled "The Impact of Digital Change on Student Learning and Mental Anguish in the COVID Era". The major goal of this chapter is to discuss the writers' research of the influence of COVID-19 on students' learning. To do so, a dataset of the different students is constructed, then analysed and displayed. The data is subjected to a variety of machine learning-based classification approaches. The support vector machine (SVM) approach is best in terms of classification accuracy, whereas the random forest (RF) method is best in terms of classification sensitivity, according to the results of the investigation.

Chapter 13 is entitled "COVID-19 Outbreak, Challenges and Possibilities to Online System of Education." Many challenges and hurdles encountered during the COVID-19 epidemic are presented in this chapter, all of which must be thoroughly investigated and handled in order to successfully adopt and assure excellent online education.

Chapter 14 is entitled "COVID pandemic: Redefining the Employment Ecosystem". The authors presented the remedies in this chapter in depth, with concrete real-life examples, to address the employment eco-system before and during the COVID epidemic. The chapter, in particular, considers factors including the work environment, human resource policy, and organizational culture. This is because an eco-system is made up of a variety of components, each of which serves a distinct purpose.

Chapter 15 is entitled "The Impact of Digital Transformation on the Story Writing Skills in COVID Era". The authors presented this pre-experimental study to see how Web-centered education in Google Classroom affected 5th grade students.

Chapter 16 is entitled "Unlock: Boon or Bane for Mental Health?" The current study attempted to investigate the mental stability of the overall population in India during the prevailing unlocking stage of the COVID-19 outbreak. Furthermore, it is intended to examine the mental health hardship in different age groups as well as the impact of psychological emotional turmoil, learning loss, financial loss, COVID exposure, and associated. The research was conducted on 200 participants from Unlock stage 3.0 to Unlock stage 8.0.

The editors are grateful to Almighty God for granting them the ability to finish this book. Completing a book is a difficult process that takes many hours, months, or even years to complete. We can speak to the fact that we worked closely with publishers, editors, and authors throughout that period. We are grateful to Nova U.S.A., our beloved Managing Editor, for believing in us and providing us with the opportunity to edit this book. Your assistance has been invaluable to us from the submission of the proposal to the conclusion of the project.

The Nova U.S.A Group deserves our gratitude. We wanted to express our gratitude to all of the authors who contributed to this book project; we were unable to accept many good chapters due to scope and quality, but we are confident that the work included in this book will be useful to young researchers and industry entrants in the fields of education, social, economic, digital transformation, IoT, artificial intelligence, and machine learning.

Our book will assist them in resolving their planned approach and provide a boost in resolving their current difficulty. We'd love to hear what you have to say about this book. Though great care was taken in selecting the chapters and writers' work, which was closely watched and amended with rigorous peer review, the evaluations and feedback of the authors or readers will be extremely helpful to us in ensuring that their concerns are addressed in our upcoming volumes. We recommend that you get this book for your institution's library and research lab, and that you use the cutting-edge technological exploration information throughout the book.

Chapter 1

Assessing the Impacts, Trends and Responses to COVID-19 in Developing Countries Using Machine Learning Techniques

Digvijay Pandey[1,*], Nwamaka U. Okafor[2], Jaspreet Singh[3], Chinwe Sam-Amobi[4], Iqra Anwer[5], Amir Shamim Shiekh[3], Muhammad Azhar[6], Vinay Kumar Nassa[7] and A. Shaji George[8]

[1]Department of Technical Education Kanpur, Kanpur, Uttar Pradesh, India
[2]School of Electrical and Electronic Engineering, University College Dublin, Dublin, Ireland
[3]Department of Social Work, Central University of Jammu, Jammu, India
[4]Department of Architecture, University of Nigeria, Nsukka, Nigeria
[5]University of Veterinary and Animal Sciences Lahore, Lahore, Pakistan
[6]Iqra University Karachi, Karachi, Pakistan
[7]Department at Sharad Institute of Technology College of Engineering, Ichalkaranji, Kolhapur, Maharashtra, India
[8]Department of Information and Communication Technology, Crown University, Int'l. Chartered Inc. (CUICI) Argentina Campus, South America

Abstract

On March 11, 2020, the World Health Organization (WHO) declared the novel coronavirus, otherwise known as COVID-19, a pandemic. Originating from Wuhan in Hubei province, China in December 2019, the pandemic has spread to more than 200 countries across the globe. The number of people infected globally by the disease has risen to more than

* Corresponding Author's Email: digit11011989@gmail.com.

In: An Interdisciplinary Approach in the Post-COVID-19 Pandemic Era
Editors: D. Pandey, B. Kumar Pandey, N. Sindhwani et al.
ISBN: 979-8-88697-052-4

2.2 million, and over 154,219 people have died as a result of this disease as of April 18, 2020. Although China, the epicentre of the pandemic, has in recent weeks recorded only a few new cases of the disease and very few deaths, The number of cases and deaths has continued to upsurge in mostly developed nations such as the United States of America (USA), Spain, Italy, France, and Germany. There is, however, speculation that developing countries, many of which have continued to record lower cases of the disease, will likely see a rise in the near future. In this chapter, we apply machine learning models for the assessment of the trend and spread of the virus in Ethiopia, Nigeria, Pakistan, and India. The analysis showed a similar trend in the rise of the epidemic in these countries and an increase in the number of cases forecast for the near future in all four countries. We evaluated the social, cultural, and economic impacts of this pandemic on these countries and how the disease is impacting the mental well-being of the masses. Finally, we analysed how government response policies have affected the spread of the pandemic and the contributions of social media to the dissemination of vital information to facilitate the containment of the global spread of the virus.

Introduction

Advances in Artificial Intelligence (AI) applications such as Machine Learning (ML), data analytics, deep learning, and facial recognition [1, 2, 3, 4] can be used for diagnosis, contact tracing, and in the development of treatment vaccines. ML has proven to be efficient in identifying and forecasting risks in diverse areas, including medical, environmental, and business fields. These technologies can no doubt be harnessed to aid in the control of the coronavirus disease and in reducing or totally curbing its impacts. Although not much research has been done on the application of ML techniques for the prediction of the spread of COVID-19, early statistics, however, present some risk factors as determinants of how likely an individual is to contract the virus. These factors include age, pre-existing conditions, social habits and hygiene, location and climate, frequency and number of human interactions, as well as social-economic and cultural factors. Machine learning methods could leverage on these factors and the current number of reported cases, recovered patients as well as adherence to other preventive measures such as social distancing, wearing of face masks and frequent washing of hands to predict the future spread and impact of the pandemic. Researchers have applied machine learning to develop an initial vulnerability index for COVID-19 [5], and have also demonstrated how machine learning could be used to predict the likelihood of a patient developing Acute Respiratory Distress Syndrome (ARDS) and a risk of mortality, just by knowledge of initial symptoms.

AI will no doubt help to provide insight into the future trend of the disease and help countries to take proactive measures to protect their citizens from the infection of this disease. While the pandemic has affected all aspects of human life, threatening the most vulnerable members of the world population, particularly in developing countries, it has also continued to impact the mental health and well-being of individuals. The current COVID-19 era has witnessed an upsurge in cases of depression, stress, anxiety, insomnia, fear, emotional distress, substance use, loneliness, and other psychological disorders. An effective psychological crisis intervention would therefore play a vital role in the overall deployment of disease control.

In this study, the authors analysed and compared the trend and spread of COVID-19 in four developing countries, including Ethiopia, Nigeria, Pakistan, and India, using machine learning models. Such a model has been previously proposed for use within a wider system for robust and effective data analysis [6]. The response of the government of these countries to containing the spread of the pandemic is analysed in relation to the trend and spread of the disease in these countries. The socio-cultural impact of the disease as well as the health and mental impact on the people in these countries are critically evaluated. Furthermore, the authors examined how social media could be a vital force in the dissemination of information necessary to contain the disease and help people maintain connections that would help them survive this unprecedented time. The impact of this research includes:

- develop suitable models capable of accurately forecasting the trend of COVID-19 in developing countries.
- present COVID-19's socioeconomic, cultural, and mental health impacts in Ethiopia, Nigeria, Pakistan, and India
- Compare the trend and spread of the pandemic in the countries under review and analyse the implications of their government's responses to the situation.

The remaining parts of this paper are structured as follows: Section 2 presents the state of the art with respect to the application of machine learning in the analysis and forecasting of the global spread of epidemics. Section 3 deals with the analysis of the trends and spread of COVID-19 in Ethiopia,

Nigeria, Pakistan, and India. Section 4 x-rays the social, cultural, and economic impact of pandemics in developing countries; section 5 analysed the impact of COVID-19 on the mental health and well-being of the masses and the importance of social media in the containment of the spread of the disease through the dissemination of vital and timely information. Results and performance evaluation of the machine learning model in the analysis of COVID-19 data are presented in Section 6, while Section 7 contains conclusions and recommendations.

1.1. State of the Art

First appearing in Wuhan, China in late 2019, the corona virus diseases is an illness that attacks the lungs and airwaves with symptoms similar to cold and flu. It can infect anyone irrespective of age and gender with the most vulnerable being the elderly and people with pre-existing conditions including those with diabetes, cardiovascular diseases, immunosuppressive conditions. The virus exhibits a high rate of person to person transmission [7, 8]. Strategies for transmission can be through droplets, sniffling and cough with patients exhibiting symptoms as fever, dry cough, difficulty in breathing, aches, pains, sore throat, headache and pneumonia. This ailment has root in group of pathogen that causes extreme intense respiratory disorder (SARS) [9, 10] which can be reached effectively by someone else. Infection rates increased rapidly and were controlled by a lockdown in most countries. In Nigeria, the government is taking active measures to minimize the spread of the disease and flatten the curve, they have taken measures such as city lock down, shut down, limitation of developments, prohibition of open assembly of over 20 people in attendance. City lockdown is a preventive activity plan executed to overcome the spread of the pandemic.

The Nigerian government is currently actively concentrating on preventive activities to fight the spread of the viral infection. According to the authors in [11], there is no definite information on the efforts of the federal government in providing food to assist its citizens during the lockdown.

The outbreak of covid-19 is developing into major international crises, raising concerns to many world leaders as the epidemic impacts almost every sector of the economy. Countries have began to shut their doors, banning international flights in and out of the country. In India, the first case was reported by the end of January 2020 and the number has increased to over 14000 by April 18 2020. India is a continent where weather and geography is variable; it is believed that these factors will affect the spread of the disease. Other variable are also important like education, financial status, so the disease rate, the ex- tent of effect will likewise be unique. Pakistan has told the direction of most countries in taking measures to curtail the spread of the pandemic. The country adopted a total lockdown approach starting from 23rd march 2020 with schools, colleges and universities haven earlier shut down since March 15 2020. The circumstance is been checked on regular routine by the bureaucratic and common place governments. The testing kits are being imported and an ever increasing number of tests are directed as time passes. Be that as it may, the quantity of tests is still very low. When of this composition, a large portion of the nation is completely bolted out. Pakistan has four issues which make them more defenseless than others including the absence of quality health care infrastructures and low economic status [12].

To help the medical institutions and government to better understand the pandemic and take appropriate action to eliminate it, artificial intelligence applications are used to analyse its trend and spread and possibly make predictions for the future direction of the pandemic. [13, 14]. Research is underway to identify drugs that were developed to fight other diseases and which could now be directed towards the fight against the infection of COVID-19. By studying the molecular composition and setup of existing drugs with AI, researchers hope to identify which one might have the capability of disrupting the operation of COVID-19.

1.2. Analysis of the Trend and Spread of COVID-19 in Selected Countries

The data used in this research is the COVID-19 data for Ethiopia, Nigeria, Pakistan and India as published in [15] and on the website of the World Health Organization [16]. Two months data from February 26 2020 till April 18 2020 were collected covering the number of confirmed daily cases, the number of daily recovered cases and the number of deaths recorded each day. A total of 57 data points were recorded. The data was first analyzed to obtain the current situation of COVID-19 in the selected countries. The trend shows a cumulative daily increase in the number of reported cases for all the countries. As at April 18 2020, India has had 14378 of her citizens tested positive for the corona virus disease out of which 480 have died, Pakistan had 7638 tested positive and 143 deaths at this date. Nigeria had 538 confirmed positive cases and 17 deaths while Ethiopia trail with 96 confirmed cases and 3 deaths as at April 18 2020.

Analysis to identify any similarity in the trend of spread of the diseases among the four countries was also performed and a fore- cast of the number of confirmed cases for the countries was done. The forecast was done with a 95% prediction interval by creating a base model with no tweak- ing of the seasonality parameters and additional regressors. All implementation of this work was done using Jupyter notebook which is shipped with Anaconda python distribution [17]. All tables and cells data were handled using pandas data frame. Visualizations were achieved by means of mat plot lib, sea born and plotly libraries and future forecasts were achieved using prophet.

1.3. Social, Cultural and Economic Impact of COVID-19 in Developing Countries

The COVID-19 outbreak has affected all aspects of human life and is threatening the most vulnerable parts of the world population, particularly in developing countries. Unlike most outbreaks whose beginning is traced to the poor living in slums and squalor, COVID-19 has spread in developing countries through the wealthy and rich. It was essentially imported into these countries by wealthy businessmen, politicians, and the well-connected; it thus began at the top.This class of people also have access to the best medical care in their countries and the world at large, yet the COVID-19 early deaths are from this class. The homeless, urban poor living in slums and squalor, people with disabilities, the elderly, and the vast population in developing countries without access to drinking water or power who survive on subsistence living are the people most at risk and vulnerable.We have reviewed the socio-cultural implications of the COVID-19 pandemic through extensive literature review and are assessing the impact and responses of the people to safety and survival considering the lock-down, social distancing, quarantine and general restriction to movement, recommendations of WHO, CDC and governments.

In developing countries like Ethiopia, Nigeria, Pakistan, and India, where there is progressive inequality, prevalence of poverty, lack of security, and discrimination of the political class towards the governed, the dangers, hardship, and desperation expected from the directives of social distancing, lockdown, and quarantine affect most of the population. In the control of the pandemic, risk communication plays a significant role by providing information that people can use to take protective and preventive action. The dissemination of basic information can rapidly and significantly reduce disease transmission; this can be seen in the reduction of COVID-19 transmission in countries that imposed early movement restrictions.In developing countries where the mistrust of the government due to high levels of corruption and social division is significant, people tend to believe rumours and misinformation as these conform to their pre-existing beliefs and anxieties [18]. This is a major challenge to the protection of the people and to controlling the spread of the disease.

Observations showed that normal communication strategies employed by the government to reach the underprivileged, the poorly educated, and the ill-informed in most developing countries, such as India, Nigeria, Ethiopia, and Pakistan, proved inadequate. Studies have identified factors such as literacy, cultural sensitivities, and reliance on oral versus written traditions that influence the way people react to communication. To be effective, risk communication must be clear, simple, and timely and delivered in a manner that will be acceptable and understood by most of the population. This will depend a lot on who delivers the message. Religious organizations, community-based organizations, and traditional rulers may form the foundation for effective risk communication. As this will be more realistic, emphatic and will address past community experiences, histories and perceptions, the communication has to be region specific. The current actions from governments of various countries, which are mostly based on recommendations from WHO, different strategies and control methods for curtailing interactions between infected and uninfected populations, include patient isolation, quarantine, lockdown, social distancing practices, school closures, use of personal protective equipment, and travel restrictions. As businesses close to help prevent the transmission of COVID-19, financial concerns and job losses are two of the first human impacts of the virus.

Strategies such as social distancing and lockdown have proved very hard in most developing countries, including Ethiopia, Nigeria, Pakistan, and India, where family and interpersonal contact is a major aspect of daily life. Practicing social distancing by especially avoiding gatherings at ceremonies, markets, mosques, and churches, has brought both emotional and psychological distress. Despite being enshrined in international health regulations [19], social distancing remains a novel phenomenon to social beings like humans. The challenges in implementing it have been enormous in non-industrialized economies, particularly among the labour class, which relies on their daily wages to put food on the table for their families.As a result, social isolation through staying at home is not an option for them, especially in the absence of adequate social security programmes and palliatives.

Socializing is unavoidable in small, poorly ventilated houses with shared washrooms and narrow streets.

Though social distancing is a major challenge in most of the developing countries, awareness and implementation must be contextualized to suit local conditions and cultural preferences. How can it be explained to parents who engage in retail sales and have to walk several kilo- meters daily selling their wares, in-order to stop their families from starvation. Many families do not understand why schools had to close believing that their children are safer in school.

1.4. Impact of Social Media on Containment of Pandemics

During the previous pandemics like the 1918 flu, critical information regarding the virus was shared through postal workers, Boy Scouts, and teachers [20]. Today we live in a society which has been described as "The Net- work Society," a society which has challenged the concept of time and space. In a time like this, when the world has been on lockdown due to a global pandemic, social media has defied geographical boundaries and had kept the world connected. Social media has contributed to educating the masses on the health implications of COVID-19, thereby saving lives. Undoubtedly social media acts as a panacea to spread awareness about the corona virus but the massive access to social media sometimes becomes a threat when false information often regarded as Corona virus Infodemic is spread via social media [21, 22, 23].

Presently social media platforms and government institutions are working together to curb the Infodemic and use social media as a tool to help flatten the COVID-19 curve. Despite such steps social media users are spreading wrong information about the virus which has proven to be a threat during this crucial time. Information related to corona virus is being posted regularly over social media platforms including statistics and the necessary pre- cautions. Celebrities including professional footballers, athletes, artists and influencers have also leveragedsocial media to reach out to people, urging them to take every- thing they see about COVID-19 seriously and follow suit, this has proven to be very effective as most people particularly the poor rural dwellers in developing regions have greater trust on these category of people.

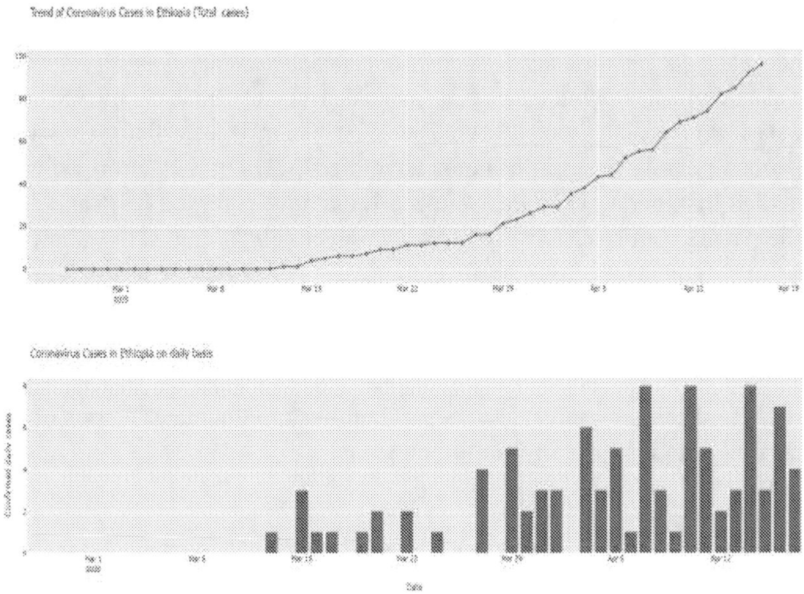

Figure 1. Trend of COVID-19 cases in Ethiopia.

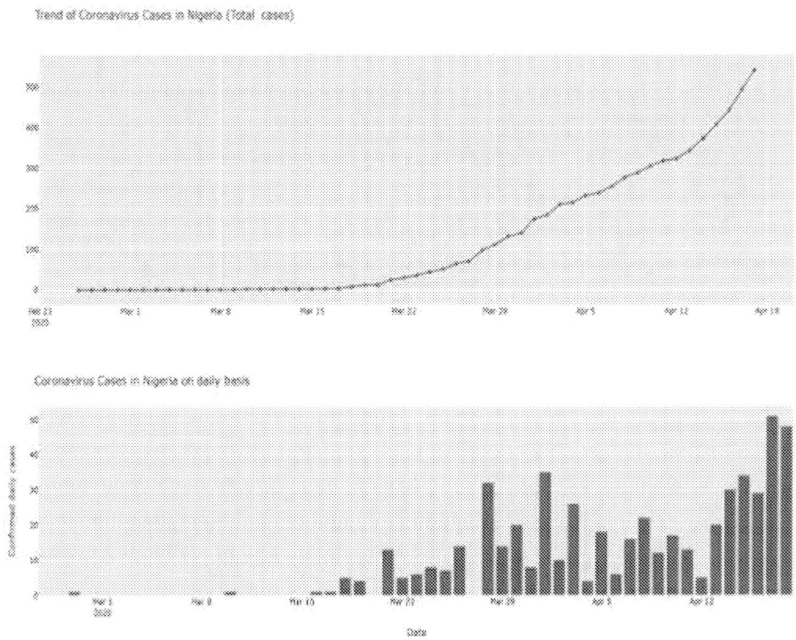

Figure 2. Trend of COVID-19 cases in Nigeria.

1.5. Results and Model Evaluation

The pandemic has continued on a forward trajectory in Ethiopia, Nigeria, Pakistan, and India. From March 26, 2020 to April 18, 2020, 3% of the confirmed positive cases in Ethiopia have died, 3.2% of the confirmed cases have died in Nigeria, while Pakistan and India have recorded 1.9% and 3.3% of their total confirmed positive cases, respectively. Figures 1–4 show the trend of the pandemic in the four countries under review. On a day-to-day basis, all four countries have recorded an inconsistent trend, alternating between a rise and a fall. The authors believe that the lack of testing kits, the unwillingness of individuals to avail themselves for testing, and the concealing of suspected cases by infected persons' families are the major causes of this fluctuating trend in the daily confirmed cases.

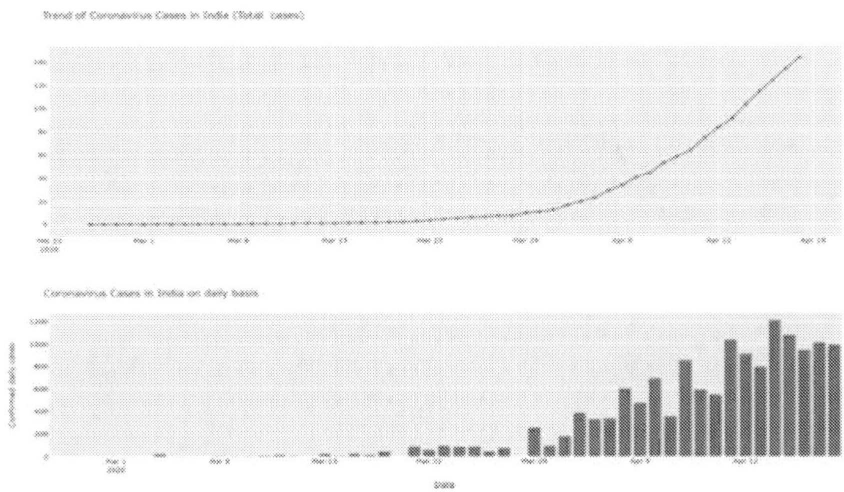

Figure 3. Trend of COVID-19 cases in Pakistan.

Figure 5 presents a comparison on the trend of the pandemic in the four countries. From the legend, the bright red region represents the lowest cases, the darker regions represent higher positive cases, with dark blue representing the highest cases. In this graph, Ethiopia and Nigeria have their bars in red, this implies that they have recorded cases which are at the bottom of the graph (lower cases, below 1k). The graph for Pakistan starts with bright red colour and gradually moved to darker red implying a continuous increase in the number of confirmed positive cases. Following the same trend, India started on the bright red side and gradually moves to the darker red up to blue, this also shows a continuous increase in the num- ber of confirmed cases. India remains on top of the list, recording a total of 14,378 positive cases within the space of 7 weeks and 2 days (from Feb 26 till April 18) out of which 480 people have died While Ethiopia trails the list with a total of 96 positive cases and 3 deaths within this period. Although India had recorded a relatively high positive cases, the number of deaths have remained relatively low (3.3% death rate from the total positive cases) whereas Ethiopia with a relative low positive cases had a 3% death rate within the period under review. See figure 6 for the plot showing the trends, including the number of confirmed cases, number of recovered cases and the number of deaths recorded in each country.

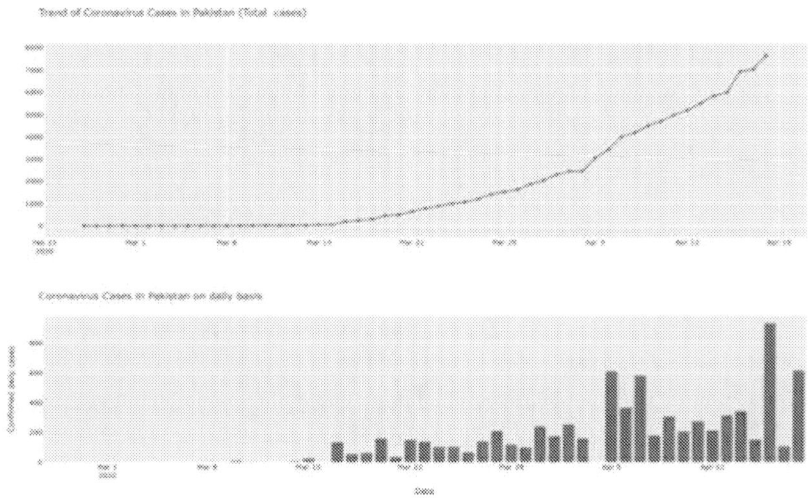

Figure 4. Trend of COVID-19 cases in India.

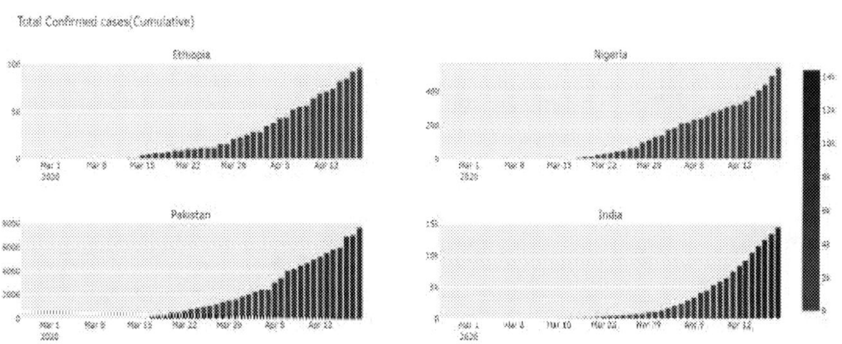

Figure 5. Comparison of the trend of COVID-19 cases in Ethiopia, Nigeria, Pakistan and India.

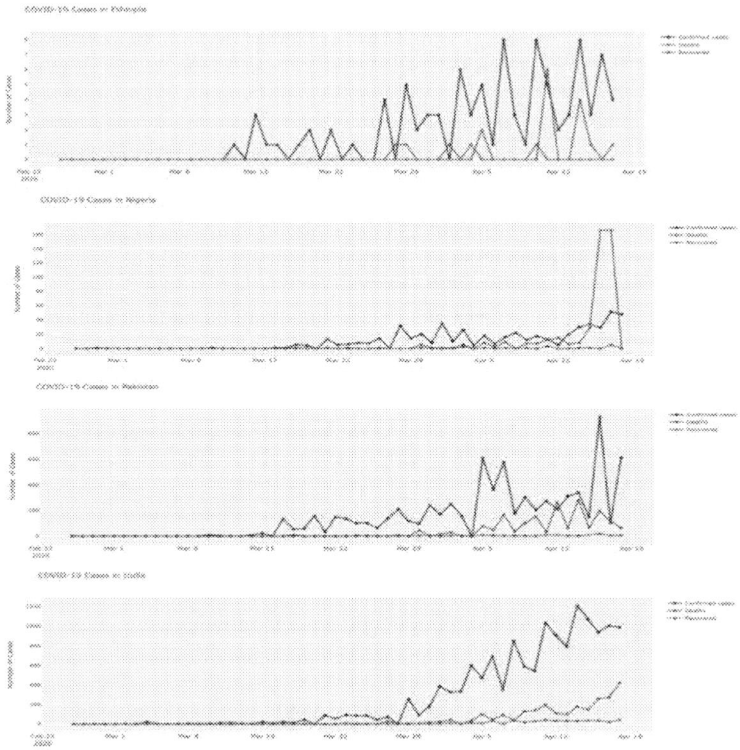

Figure 6. No of confirmed, recovered and death cases of COVID-19 in Ethiopia, Nigeria, Pakistan and India.

In Figures 7 and 8, we present the plots representing the ML- based forecast for the trends of COVID-19 in Ethiopia and India (the result for Nigeria and Pakistan is included in the supplementary material), In particular, the expected number of positive cases in the selected countries for the next 2 weeks (period of 26 April 2020 to 01 May 2020) is forecast/predicted. The black dots on the upper plots are the original values from the historical data, the blue line represent the forecast values with the upper and lower tolerance. The trends of cases in each day of the week are also shown (lower plots). Ethiopia is seen to be recording high cases on Tuesdays and Fridays whereas India have her lower cases on these days and peaks on Wednesdays. The forecasts presented similar trends with the historical trend of the past weeks. The forecasts predicts that all the four countries under review are likely to face more challenging days ahead with regards to the spread of the disease if appropriate measures are not strictly put in place and adhered to in view of containing the pandemic.

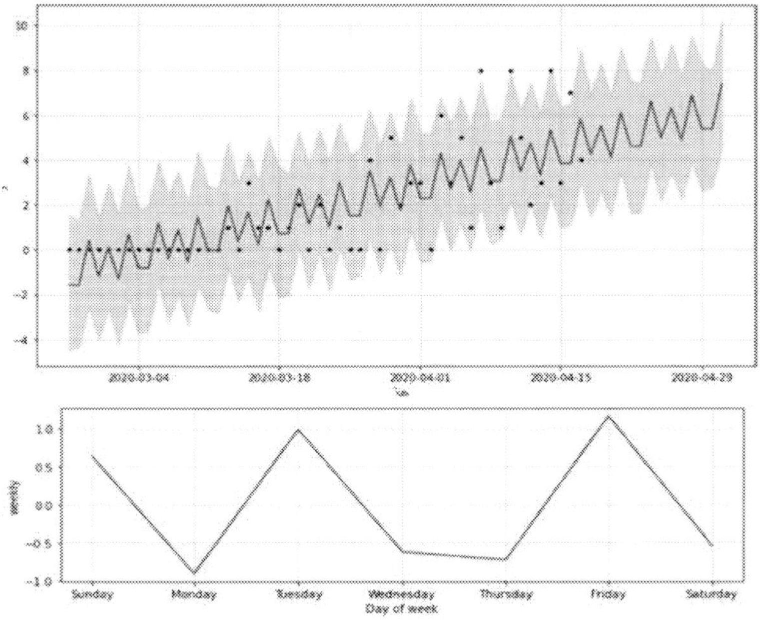

Figure 7. Future forecast of COVID-19 cases in Ethiopia.

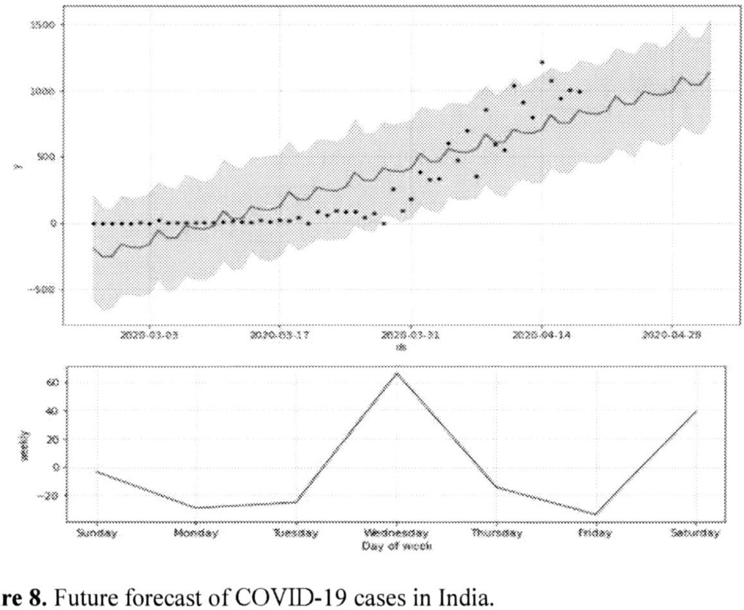

Figure 8. Future forecast of COVID-19 cases in India.

Conclusion and Recommendations

While the government of Ethiopia, Nigeria, Pakistan and India have continued to adopt appropriate measures to curtail the spread of COVID-19 in their various countries, The virus, however have maintained an increasing rate of spread in all four countries with the number of confirmed positive cases and the number of deaths rising rapidly. The authors therefore recommend that governments should maintain measures aimed at flatten the curve of the spread in order not to overwhelm the limited health facilities which were identified as issues common to all the four countries investigated. The analysis performed in this study showed similarity in the rising spread of the virus wile predicting further rise in the coming days in all the four countries. It is therefore imperative to ensure that the various measures currently in place by the governments which are aimed at reducing the spread of the disease are strictly adhered to by all citizens. During the period of investigation, the authors identified that all four countries consistently records high number of cases on certain days of the week. For instance, during the 7 weeks study period, Ethiopia was found to record high number of positive cases on Fridays while Nigeria and India had high number of cases recorded on Wednesdays. The authors could not establish the cause of this during this re- search, however, future research could look to determine the reason for higher positive cases on certain days of the week in these countries. The study have also identified an increase in mental-health related challenges duringthe COVID-19 period [24, 25]. Several issues have been identified in this study as contributing factors to such mental-health challenges at this time and these include frustration arising as a result of people being unable to make ends meet and provide food for their family due to certain measures by the government including lockdown, isolation,closure of market and businesses, restrictions to public gathering and travels [26]. To help alleviate these problems, the authors recommends that governments should ensure that palliatives are provided to the citizens to help cushion the effects of the lockdown and other containment measures.

References

[1] Juneja, S., Juneja, A., & Anand, R. (2019, April). *Reliability modeling for embedded system environment compared to available software reliability growth models*. In 2019 International Conference on Automation, Computational and Technology Management (ICACTM) (pp. 379-382). IEEE.

[2] Meivel, S., Sindhwani, N., Anand, R., Pandey, D., Alnuaim, A. A., Altheneyan, A. S., ... & Lelisho, M. E. (2022). Mask Detection and Social Distance Identification Using Internet of Things and Faster R-CNN Algorithm. *Computational Intelligence and Neuroscience*, 2022.

[3] Sindhwani, N., Anand, R., Meivel, S., Shukla, R., Yadav, M. P., & Yadav, V. (2021). Performance Analysis of Deep Neural Networks Using Computer Vision. *machine learning*, 15, 17.

[4] Pandey, B. K., Pandey, D., Wairya, S., & Agarwal, G. (2021). An advanced morphological component analysis, steganography, and deep learning-based system to transmit secure textual data. *International Journal of Distributed Artificial Intelligence (IJDAI)*, 13(2), 40-62.

[5] Schmitt M., *"Fight COVID-19 with machine learning,"* Medium, 2020. [Online]. Available: https://towardsdatascience.com/fight-covid-19-with- machine-learning-1d1106192d84. [Accessed: 27- Apr- 2020].

[6] Okafor N. and D. Delaney, "Considerations for system design in IoT-based autonomous ecological sensing," *Procedia Computer Science*, vol. 155, pp. 258-267, 2019. Available: 10.1016/j.procs.2019.08.037.

[7] Abere O. J., "Survival Analysis of Novel Corona Virus (2019-Ncov) Using Nelson AalenSurvival Es- timate," *International Journal 0f Business Education And Management Studies*, vol. 3, no. 1, pp. 30-40, 2020.

[8] Singh J. and J. Singh, "COVID-19 and its impacts on the society," *Electronic Research Journal of Social Sciences and Humanities*, Vol 2: Issue I, 2020.

[9] Jain, S., Kumar, M., Sindhwani, N., & Singh, P. (2021, September). *SARS-Cov-2 detection using Deep Learning Techniques on the basis of Clinical Reports*. In 2021 9th International Conference on Reliability, Infocom Technologies and Optimization (Trends and Future Directions) (ICRITO) (pp. 1-5). IEEE.

[10] Jain, S., Sindhwani, N., Anand, R., & Kannan, R. (2022). COVID Detection Using Chest X-Ray and Transfer Learning. In *International Conference on Intelligent Systems Design and Applications* (pp. 933-943). Springer, Cham.

[11] Nseobot I., D. Hamid, D. Elyassami, A. Effiong, U. Ette, M. Ahmed-Soomro. *COVID-19 CityLocked Down: Implications on Human Welfare in Developing Countries*, 2020.

[12] Bloukh S. H., A. A.Shaikh, H. M.Pathan, Z. Edis. Prevalence of COVID-19: A Look Behind the Scenes. Preprints 2020, 2020040179 (doi: 10.20944/preprints202004.0179.v2).

[13] Anand, R., Sindhwani, N., & Saini, A. (2021). Emerging Technologies for COVID-19. Enabling Healthcare 4.0 for Pandemics: A Roadmap Using AI, Machine Learning, IoT and Cognitive Technologies, 163-188.

[14] Babu, S. Z. D., Pandey, D., & Sheik, I. (2020). Acts Of COVID19. *International Journal of Advanced Engineering*, 3, 2457-0397.

[15] "Coronavirus Update (Live): 3,065,176 Cases and 211,631 Deaths from COVID-19 Virus Pandemic – Worldometer," Worldometers.info, 2020. [Online]. Available: https://www.worldometers.info/coronavirus/. [Ac- cessed: 28- Apr- 2020].

[16] *"WHO COVID-19 Dashboard,"* COVID19.who.int, 2020. [Online]. Available: https://covid19.who.int/. [Accessed: 28- Apr- 2020].

[17] *"The World's Most Popular Data Science Platform Anaconda,"* Anaconda, 2020. [Online]. Avail- able: https://www.anaconda.com/.

[18] Greenhill K. and B. Oppenheim, "Rumor Has It: The Adoption of Unverified Information in Conflict Zones," *International Studies Quarterly*, vol. 61, no. 3, pp. 660-676, 2017. Available: 10.1093/isq/sqx015.

[19] Nuttall I., "International Health Regulations(2005): taking stock," *Bulletin of the World Health Organization*, vol. 92, no. 5, pp. 310-310, 2014. Available: 10.2471/ blt.14.138990.

[20] *"The role of social media during a pandemic Khoros,"* Online Community Social Media Management Software — Khoros, 2020. [Online]. Available: https://khoros.com/blog/social-medias- role-during-covid-19. [Accessed: 28- Apr-2020].

[21] Hussain Wajahat, "Role of Social Media in COVID- 19 Pandemic," *The International Journal of Frontier Sciences*, vol. 4, no. 2, 2020. Available: 10.37978/ti- jfs.v4i2.144.

[22] Khongsai, L., Anal, T. S. S. C., Rapheileng, A. S., Tomba Singh Kh, Md Kheiruddin Shah, Digvijay Pandey. Combating the Spread of COVID-19 Through Community Participation. *Glob. Soc. Welf.* (2020). https://doi.org/10.1007/s40609-020-00174-4.

[23] Kassaw, C., Pandey, D. The Current Mental Health Crisis of COVID-19 Pandemic Among Communities Living in Gedeo Zone Dilla, SNNP, Ethiopia, April 2020. *J. Psychosoc. Rehabil. Ment. Health* (2020). https://doi.org/10.1007/s40737-020-00192-7.

[24] Rather, R. A., Islam, T., Rehman, I. U., & Pandey, D. (2021). Development of vaccine against coronavirus disease 2019 (COVID-19) In India. *Asian Journal of Advances in Medical Science*, 13-21.

[25] Bakshi, G., Shukla, R., Yadav, V., Dahiya, A., Anand, R., Sindhwani, N., & Singh, H. (2021). An Optimized Approach for Feature Extraction in Multi-Relational Statistical Learning. *Journal of Scientific and Industrial Research (JSIR)*, 80(6), 537-542.

[26] Zargar, S. A., Islam, T., Rehman, I. U., & Pandey, D. (2021). Use of Cluster Analysis To Monitor Novel Corona Virus (COVID-19) Infections In India. *Asian Journal of Advances in Medical Science*, 1-7.

Chapter 2

Students' Behaviour in Facing Cultural Arts to Support Child Friendly School (CFS): Preserving Local Culture at Schools during the COVID-19 Crisis in Gayo, Aceh

Siti Supeni[1,*], Oktiana Handini[2,†] and Luqman Al Hakim[3,‡]

[1]Civics Education at Slamet Riyadi University, Surakarta, Central Java, Indonesia
[2]Elementary School Teacher Education at Slamet Riyadi University, Surakarta, Central Java, Indonesia
[3]English Language Education at Slamet Riyadi University, Surakarta, Central Java, Indonesia

Abstract

The chapter aims to show the students' activities in preserving local cultural arts at school to plan, implement, and evaluate during the COVID-19 crisis. The type of the research is descriptive qualitative. The subject of the research is 78 elementary school students in Gayo, Aceh, Indonesia. They are from a variety of elementary schools. The sources of the data are informants and documents. The informants are the elementary students. The document is a lesson plan for teachers. The techniques for collecting the data are a questionnaire and document analysis. We distributed the questionnaire after a focus group discussion occurred in Gayo. The technique for analysing the data used an

* Corresponding Author's Email: sitisupeni56.unisri@gmail.com.
† Corresponding Author's Email: handinioktiana7@gmail.com.
‡ Corresponding Author's Email: luqmanalhakim@unisri.ac.id.

In: An Interdisciplinary Approach in the Post-COVID-19 Pandemic Era
Editors: D. Pandey, B. Kumar Pandey, N. Sindhwani et al.
ISBN: 979-8-88697-052-4

interactive analysis that comprised data reduction, data display, and conclusion. The result of the research is that the students actively took part, cooperated, and felt excited in conducting the local cultural arts agenda. Therefore, they could have achievement in this field.

Keywords: child friendly school (CFS), behavior, local culture

1. Introduction

School is a place where students get knowledge, help, and rights and grow their potency and interest [1]. Positive schools have many criteria: promoting teaching strategies without pressure and punishment; facilitating peer support and collaboration; preventing bullying and abuse; and giving everyone a chance to learn and develop equally. The school's criteria have also fulfilled the basic facilities to support a safe, caring, and child-friendly school [2]. Schools also have supporting and taking part factors actively from principals, teachers, students, and the community in order to give a standard learning environment to support child-friendly schools [3]. School management, an effective learning environment, interaction among teachers and students, and community participation are activities to realize a child-friendly school [4]. Besides, to create CFS, schools need to facilitate students' creativity to learn strategy and the approach of the school environment [5]. Pedagogical factors are also a part of policy to support successful CFS [6].

The basic education level has the most urgent role in the progress of a nation. At this level, we lay the foundation for knowledge and character development in students. One form of strategy to fulfil children's education rights in schools is the creation of a child-friendly school/madrasah model. The existence of this friendly school or madrasah becomes very urgent if they associate it with the current state of education, which is prone to bullying or bullying activities. The development of the concept of a child-friendly school is not only focused on managerial and institutional aspects but more broadly on several aspects, one of which is in the aspect of regional culture-based character, which is especially related to the participation of teachers, media, methods, and learning resources.

Implementing CFS must not always be child-friendly, i.e., school regulations that convey the importance of routine in schools or discipline for children having learning and/or behavioural difficulties [7]. Safe, caring, and CFS have a problem with inadequate respect to CFS status for most basic facilities in schools [2]. The lack of a person to manage CFS, the attitude of teachers, a lack of awareness of students, and a lack of facilities are the problems in CFS [3]. CFS policy not only ignores local cultural conditions but also eliminates the importance of local pedagogical forms; therefore, it needs a more nuanced approach that identifies and accommodates various forms of pedagogical practice [6].

It is in this context that there is a discussion regarding child-friendly school activities (SRA) in regional culture-based character education in elementary schools. Thus, education is child-friendly if, by looking at the conditions of a safe, comfortable, and loving learning environment at school, the relationship that exists with a sense of love and affection between students and teachers, parents, and their peers is very influential in shaping the character of a child. So, it needs a framework of rights-based tools to help children's rights and to provide good quality education [8] in the COVID crisis. Using the UNICEF framework for rights-based child-friendly educational systems to analyse how far the schools' barriers resulted in the unavailability of teaching and learning resources and schools' poor condition because of the rejection of the convention on children's rights [9].

The reality of CFS needs to be evaluated based on findings at schools. Gender of students, grade level, and socioeconomic school levels are factors that make CFS features of schools change. Therefore, it needs to create a sense of awareness for the CFS concept [10]. There are some conditions of CFS, i.e., the disappearance of resources for teaching and learning, ignorance of the convention on children's rights, and students do not know their rights; therefore, it needs the involvement of government and stakeholders at schools [9]. There is a difference of opinion between pupils and teachers about child-friendly schools and the confusing concept of CFS; therefore, it needs to employ the concept carefully [7]. Unfriendly, noisy conditions at schools cause the schools to not meet Child-Friendly Schools criteria [11].

Support for students is one of the components of successful CFS implementation [3]. Strong relationships between teachers, students, and parents are a way of improving the quality of education besides teachers and principals' activities, curriculum, inside and outside school and family environment, and character education [12]. Giving students opportunities to learn and grow, as well as facilitating collaboration, are examples of positive schooling [1]. Students' participation is one of six important principles of CFS based on the context of education in Indonesia [13]. In this research, the authors intend to reveal research findings about the students' contribution to schools in terms of cultural arts to support child-friendly schools in COVID time.

1.1. Overview of Child Friendly School

The concept of child-friendly schools (CFS) emerged in the 1990s in Asia. This is to respond to the rights-based, child-friendly conceptual framework developed by UNICEF. In a right-based framework, schools must have basic characteristics. First, they have to be a child-seeking school. Schools should actively collect excluded children so that the schools enrol them and include them in learning. The schools should treat all children as entitled subjects and the state as a duty and obligation to attain those rights. Second, the school must be a school that is central to the child, acting in the best interests of the child in a way that will fulfil the child's full potential [9]. Child-friendly schools must ensure the opportunity for every student to attain the right to education without discrimination because of disability, gender, ethnicity, type of intelligence, and background of the parents. Child-friendly schools should also examine safe, clean, and healthy school situations, care and culture, the living environment, respect the rights and protection of students from violence, discrimination, and other unfair treatment, and provide student participation in planning, policy, learning, supervision mechanisms, and complaints related to fulfilling the rights and protection of learners in education [5].

Child-Friendly Schools are schools that have the basic rights of children. Schools are considered child-friendly while providing a safe, clean, healthy, and protective environment for children [14]. Child-friendly schools are concepts that include variables such as school buildings, environmental planning, learning-teaching processes, resources and materials, teachers, principals, health, safety, democratic participation, and gender sensitivity [10].

UNICEF has developed a framework for a rights-based, child-friendly education and school system that is featured as "inclusive, healthy and protective for all children, effective with children, and engaging with families and communities-and children." In this framework,

1. School is a significant personal and social environment in a student's life. Child-friendly schools ensure every child has a safe, emotional, and psychologically safe physical environment.
2. Teachers are the single most important factor in making learning effective and efficient in inclusive classes.

3. Children are natural learners, but this capacity for learning can be eroded and sometimes destroyed. Child-friendly schools recognize, encourage, and support a child's capacity as a learner by providing a school culture, behaviour, and curriculum content that focuses on learning and learning.
4. The school's ability to be and call her child-friendly is directly related to the support, participation, and collaboration it receives from the family.
5. Child-friendly schools aim to develop a learning environment in which children are motivated and able to learn. Staff members are friendly and respectful of children and take care of all health and safety needs [8].

1.2. Culture in Gayo, Middle Aceh, Indonesia

Gayo is a city in the Middle Aceh region, well known for its culture and Islamic values. The following are the cultures in Gayo:

1. *Saman* This dance is a traditional dance in Aceh that has unique characteristics. It has interaction and cooperation activities among dancers. In addition, it has a culture of respecting the abilities of others and mutual aid, so that if there are members of the group who have difficulty performing dance movements, the other students can offer help [15].
2. *Tawar Sedenge*. Talking about a song from Aceh, there is a song that contains patriotism. This is *Tawar Sadenge.* For the gay community, this song has an essential form to this day. This song is commonly played at the official events of Takengon and every national ceremony as an interlude to the song Indonesia Raya. It plays an important role in the Gayo community because the lyrics of this song contain the spirit of nationalism that can boost the spirit of the Gayo community to achieve progress and prosperity. The lyrics and musical elements in this song prove that the song has a Gayo cultural identity, which is based on the elements of cultural identity based on beliefs, languagc, and behaviour [16].
3. *Petawaren.* This tradition is purely Gayo culture that has more colours of the Islamic dialectic and the *Petawaren* tradition in the Gayo community that has more Islamic values. So, interpreting

symbols, practices, and customary objects is according to Islamic teachings that replaced the beliefs of animism that existed before Islam [17].

4. *Peri Mestike*. This culture is one of the original values of wisdom and culture in Aceh. Gayo people are one of the ethnic groups in Indonesia that preserve, respect, and like the high values of their *Adat* and culture. To this day, they still have a firm to maintain because they believe that Gayo customs and culture can guide them in life. Forming harmony in society, shaping or building morals, and not contradicting the values of Islamic teachings are the characteristics. The internalization of Peri Mestike in the world of education is part of a moral development effort and is an effort to preserve Gayo's original values of wisdom and culture [18].
5. *Gayo Sumang Cultural System*. This cultural system contains knowledge, beliefs, values, rules, and laws that serve as a reference in the Gayo community. This paper seeks to understand the cultural system of the popular Gayo community in Sumang, which means keeping it from etiquette conventions and is contrary to Islam and customs. Implementing a spirit culture for the restoration of the character of the Gayo community is very relevant as it is spiritual and oriented on the moral of al-karmah, maintaining self-esteem, family and social dignity. Self-esteem is called mukemel, which means "to have a sense of shame." The author concludes that culture plays an important role in the culture of a society to be higher morally when one implements it in all aspects of society [19].
6. *Sengen Tradition*. This tradition has some function: to keep their children well and to spare them from bad things. It means to reduce the burden of life and to show love and affection to the child for survival and responsibility to the child. One believes and proves that this unique tradition carries positive results, such as bringing safety to children. Many people know the true meaning of love, but they keep doing it because they believe that this act of love can distract them from bad things. However, he still believes that God determines everything [20].

2. Method

The type of the research is descriptive qualitative. The subject of the research is 78 elementary school students in Gayo, Aceh, Indonesia. They are from a variety of elementary schools. The sources of the data are informants and documents. The informants are the elementary students (as I mentioned above). The document is a lesson plan for teachers. The time of the research is from April 2021 up to September 2021.

Data collection techniques [21] are carried out on related sources selectively by using considerations based on the theoretical concepts used. The technique used in this research is more *purposive sampling.*

2.1. Data Sources and Data Collection Techniques

Data collection techniques in this study include:

1. Delivery *Questionnaire:* Questionnaire via *Google Forms* to all respondents in Gayo, Middle Aceh
2. Interviews with resource people related to the object of research
3. Recording data on various aspects related to child-friendly cities and regional cultural development.
4. Studying and reviewing the literature can provide information to support this research.
5. Documenting through photo shoots of data sources such as regional cultural activities carried out by children in Surakarta.

2.2. Data Analysis

Reviews concerning the analysis in this study emphasise more on the interaction model of qualitative data analysis using a cultural approach. Analysing qualitative data from empirical data collection to get accurate results from sorting by classification and identification is essential [22, 23, 24]. The researcher chose this model because it allows them to provide more one-shots that can capture input and exposure in a summary that is data reduction and inference. Qualitative data by applying a cyclical system means that the researcher is always moving and exploring the object during the process. Several types of data sources needed in this study include resource people related to the object of research, library resources related to character education, regional cultural development, and child-friendly cities in the four provinces of the research area.

3. Results

3.1. The Students' Behaviour in Cultural Arts to Support a Child-Friendly School in a COVID crisis

The students' contribution to schools in terms of cultural arts to support child-friendly schools is divided into four aspects: planning, implementation, evaluation, and general contribution. It means that the students contribute to the four components of preserving CFS.

Table 1 shows that the four elements achieve a satisfying percentage of the planned activities of the students before doing local cultural arts activities. The 1st statement conveys that there are 92.3% (72 students) who follow the school's agenda: local cultural arts; and 7.7% of those (6 students) do not follow the agenda. In the 2nd statement, those who are excited to be trained by trainers about local cultural arts are 97.4% (76 students), and those who are not excited are 2.6% (2 students). The next statement reveals that 91% (71 students) who reveal their discipline and obey the regulation to take part actively in local cultural arts inside and outside my school, and 9% (7 students) who disobey. Last, there are 61.8% (48 students) who win the competition for local cultural arts, and 38.2% (30 students) do not win the competition.

Table 2 is implementing local cultural arts. There are 97.5% (76 students) who follow local cultural art activities based on their potency and interest, and 2.5% (2 students) who do not follow. The 2nd statement shows that those who are excited to follow local cultural arts with the trainers are 94.9% (74 students), those who are not, are 5.1% (4 students). Then, those who are proud of being able to speak the local language, dance local dances, sing local songs, and wear local uniforms are 100% (78 students). Finally, there are 92.3% (72 students) who decrease the use of broad arts (tik-tok, Korean dance, etc.), and 7.7 students (6 students) who increase it.

Table 1. Planning activities

	Statements	Percentage (Yes/no)
1	I follow school's agenda: local cultural arts	Yes (92.3%); No (7.7%)
2	I am excited to be trained by trainers about local cultural arts	Yes (97.4%); No (2.4%)
3	I am discipline and obey the regulation to take part actively in local cultural arts inside and outside my school	Yes (91%); No (9%)
4	I win the competition of local cultural arts	Yes (61.8%); No (38.2%)

Table 2. Implementation

	Statements	Percentage (Yes/no)
1	I follow local cultural art activities based on my potency and interest	Yes (97.5%); No (2.5%)
2	I am excited to follow local cultural arts with the trainers	Yes (94.9%); No (5.1%)
3	I am proud of being able to speak local language, dance local dances, sing local songs, and wear local uniform	Yes (100%); No (-%)
4	I decrease the use of abroad arts (tik-tok, Korean dance, etc.)	Yes (92.3%); No (7.7%)

In the evaluation part in Table 3, there are two statements. 97.4% (76 students) are excited about the following extracurricular (local cultural arts) inside and outside of my school: 97.4% (76 students), and 2.6% (2 students) are not. The last, those who give an assessment about implementing local cultural arts activities and competitions are 77.2% (60 students) and those who are not, get 22.8% (18 students).

In the Table 4, there are five questions about the roles of the students in preserving local cultural arts. In the 1st question, there are 75 students (96.2%) who did cultural arts to support CFS and 3 students (3.8%) did not. Then, in the 2nd question, there are 69 students (88.5%) who got support from their parents in joining competition and performance of local cultural arts, however, 9 students (11.5%) did not. Subsequently, the 3rd question is about discipline and attitude of the students in joining local cultural arts. There are 75 students (96.2%) who were disciplined and obeyed the rules in joining local cultural arts, and 3 students (3.8%) were not. In the 4th sentence, there are 48 students (61.8%) who got used to behaving politely at home and at school and 30 students (38.2%) did not. Finally, there are 29 students (37.7%) who liked local dances, 31 students (40.3%) who liked traditional songs, and 17 students (22.1%) who likes traditional games.

Table 3. Evaluation

	Statements	Percentage (Yes/no)
1	I am excited about participating in extracurricular (local cultural arts) inside and outside of my school	Yes (97.4%); No (2.6%)
2	I and my teacher give assessment about implementing local cultural arts activities and competition	Yes (77.2%); No (22.8%)

Table 4. Students' roles in cultural arts

	Statements	Percentage (Yes/no)
1	I and the trainer do cultural arts to support child friendly school (CFS)	Yes (96.2%); No (3.8%)
2	I get support from my parents in joining competition and performance of local cultural arts	Yes (88.5%); No (11.5%)
3	I am disciplined and obey the regulation in joining local cultural arts	Yes (96.2%); No (3.8%)
4	I get used to behaving politely at home and at school	Yes (61.8%); No (38.2%)
5	I like local cultural arts	Local dances (37.7%) Traditional songs (40.3%) Traditional games (22.1%)

3.2. The Inhibited Factors of Child Friendly School (CFS) in COVID Crisis

Based on the interview with the Principals of Schools in Gayo, there are some inhibited factors in CFS below:

- The development of technology is quick, and the impact of globalisation can erode cultural values and create an environment that is not conducive to children's development.
- Crimes that stalk children both psychologically and physically have more advanced modes by utilising digital technology.
- Insufficient infrastructure for arts and cultural activities. The interview with the principal, it was noted that the limited

infrastructure, especially in public elementary schools, was related to cultural arts activities such as gamelan in Java and other musical instruments originating from other research areas. However, introducing and giving traditional musical instruments is a must since the child's age is as early as possible. The change in digital-based community civilization towards the 4.0 industrial revolution, of course, has a great impact on changes in regional culture in society. The industrial revolution is based on the digital revolution, where time distance becomes very effective. People work more; the development of information is quick, as is the shift in community values.

- They are individualistic and have less humanistic value because they do not depend on direct interaction with others. Through digitization, we can meet all needs and interests. Distance is no longer a problem. There are no longer visible differences between cities, villages, or mountainous areas because humans can manipulate demographic conditions. This, of course, brings different values to digital native children who prefer instant and modern, so that they will not recognise cultural values if they do not get into the development of their lives.

4. Discussion

Over 60% of students take part actively in the school's planning. The three aspects have a combined score of more than 90%. It means that they obey the rules to do planned activities. It is in line with [13] who reveals that students' participation is one of the key successes of implementing CFS. Collaboration is one of the successful activities to support CFS; therefore, it is in line with one of the collaborative activities [1]. Management of the school to handle and facilitate the implementation of local cultural arts plays an important role in ensuring anything runs well. It is in line with [4] stating that a school's management, effective learning environment, interaction among teachers and students, and community participation are activities to realise a child-friendly school.

Fulfilling children's rights during their studies at schools is an indicator of successful child-friendly schools because these schools devote time to facilitating them to optimise their potency, interest, and skills. This research shows that schools, with the support of parents and the school's community, have fulfilled the rights of children to facilitate their activities to preserve local culture. The role of parents (those having their children study at the intended schools) and community plays an important role in increasing understanding and commitment to children's rights and improving the quality of education [4]. Genuine community involvement can improve education, community awareness, and CFS [12].

Students at the elementary school level in Gayo, Aceh, receive guidance according to their talents and interests in the arts and culture. They take part in the training by practising once a week at school and taking part in cultural arts performances regularly. Their motivation to practise is because of their own will to perform. They shared knowledge and experiences with friends in studying regional culture and taking part in cultural arts outside of school and practising cultural arts alone at home with friends to improve their local cultural arts skills. Regional cultural support facilities support these activities to learn character education for the local arts and culture that are occupied. The students have carried out regional cultural arts activities at school to shape children's character. The students did disciplinary behaviour to show the implanting formation of their character. The display of educated behaviour is as follows: academic achievement, ethical behaviour, and noble character; high learning motivation; creativity; discipline; responsibility; and showing self-character as a good citizen. This is in line with [1] that learners and facilitators play an important role in supporting each other, focusing on the advantages rather than disadvantages of creating an inclusive environment that accommodates everyone to realise CFS.

Most of the *Aceh* region is still thick with culture, or customs that are still firmly held today. The cultural system of the Gayo community is spiritually valuable and oriented towards *akhlaq al-karmah.* These cultural values shape the association of living together based on Islamic law. The cultural values in question are deliberation, trust, trust, discipline, mutual help, affection, loyalty, and competition. One knows important values in the customs and culture of the Gayo people as the principle that seeing God or the Holy Spirit as an important norm or principle of life can be called "spiritual." The Gayo custom serves to implement the law based on reality. While the law is based on the Word of *Allah* and His Messenger in the Qur'an and Hadith. According to [4], the roles of parents and the community are critical in increasing understanding and commitment to children's rights and improving educational quality.

Implementing these values in local content (*muatan lokal*) connects the needs of families and communities with the goals of national education. These subjects also provide opportunities for students to develop their abilities, which are deemed necessary by the region concerned. Therefore, the local content, especially in the Gayo Highlands, must contain characteristics of Gayo culture itself (local), skills, and noble values of local culture to raise social and environmental problems, which in the end can equip students with basic skills as provisions in life (life skills), so that they can create jobs, and students have the provision of norms and manners of noble behaviour according to the expectations of the goals of education itself. The local content, which is centred on local customs and culture, has a function to lift culture and shape the moral values of students, and also, culture is the identity of a regional clump, so maintaining it is a must, because culture contains social and cultural values and norms that Islam.

Conclusion

The development of character education based on regional arts and culture, with a variety of cultural patterns, will create pride as an Indonesian nation, with a strategic area currently focused on the concept of regional tourism development as Indonesia's potential strength. Building a strong cultured country requires human resources, especially the involvement of all elements of society to take part in development. The preparation of human resources through education is an urgent matter to be done immediately [25]. So, local governments have a big responsibility to realise child-friendly education.

So far, several elementary schools have not had access to quality improvement efforts. This is because these schools are far from the education office, which is located at the centre of the regional government. In addition, schools receive less attention and are involved in research activities carried out by universities. Regional culture, when applied, has local wisdom values as a reference in developing child-friendly schools that are under regional and cultural contexts. Re-exploring traditional values becomes urgent when the development of education that has been happening so far is technocratic, which actually damages the environment.

Elementary school students arc part of a society that is very vulnerable to the rush of unfiltered foreign culture, including cultural arts that are not part of our national identity. We carried the research out, especially at Child

Friendly Schools in four provinces, to explore the potential of their regional arts and culture based on character education.

Acknowledgments

The researchers convey deep gratitude to Allah SWT for time and opportunity. The researchers are grateful to the Ministry of Education, Culture, Research, and Higher Education for providing funding for this research. Finally, we say thank you to all the lecturers and students who supported this research.

References

[1] Thomas, S., Jose, K. A. and Kumar, P. A. "Child Friendly Schools : Challenges and Issues in Creating a Positive and Protective School Environment," in *Positive Schooling and Child Development*, S. Deb, Ed. SIngapore: Springer Singapore, 2018, pp. 233-248.

[2] Makwarela, M. C., Adu, E. O. and Mammen, K. J. "The Intervention of Safe, Caring and Child-friendly School Policies on Social Construction of Violence in South African Secondary Schools," *J. Soc. Sci.*, vol. 8923, no. 50, pp. 8-13, 2017.

[3] Silao-Navarra, E. M. A. E. "The Goals of a Child-Friendly School and the Psycho-Social Environment of the Learners in Public Secondary Schools in the Division of Butuan City, Philippines," *SMCC Teach. Educ. J.*, vol. 1, no. June, pp. 163-178, 2019.

[4] Ardiansyah, A., Hariri, H. and Sowiyah, "School-Based Management In Realizing Child-Friendly Schools In The Era Of COVID-19 Pandemic In An Indonesian School," *Int. J. Educ. Res. Soc. Sci.*, pp. 637-647, 2021.

[5] Lian, B., Kristiawan, M. and Fitriya, R. "Giving creativity room to students through the friendly school's program," *Int. J. Sci. Technol. Res.*, vol. 7, no. 7, pp. 1-7, 2018.

[6] King, E. "Implications for the child friendly schools policy within Cambodia' s cultural and primary school context," *Asia-Pacific J. Teach. Educ.*, pp. 1-14, 2019.

[7] Leino, M. "The Child Friendly School : An Idea Versus Reality," *Probl. Educ. 21st century*, vol. 29, pp. 82-88, 2011.

[8] Orkodashvili, M. "Quality education through Child-Friendly Schools: resource allocation for the protection of children's rights," *Munich Pers. RePEc Arch.*, no. 23520, 2010.

[9] Mandiudza, L. "Child-Friendly Schools," *Greener J. Educ. Res.*, vol. 3, no. 6, pp. 283-288, 2013.

[10] Çobanoğlu, F., Tuncel, Z. A. and A. Ordu, "Child-friendly Schools : An Assessment of Secondary Schools," *Univers. J. Educ. Res.*, vol. 6, no. 3, pp. 466-477, 2018.

[11] Syafrudie, H. A. and Haryo, A. "School readiness in supporting child-friendly schools reviewed from noise aspect of the classroom in a primary public school in the new normal era," *IOP Conf. Ser. Earth Environ. Sci.*, vol. 802, no. 1, 2021.

[12] Hermino, A. and Luangsithideth, V. "Community Awarenes and Child-Friendly School at the District Nonghet, Northern of Lao PDR," *Asian Soc. Sci.*, vol. 13, no. 11, pp. 109-116, 2017.

[13] Fitriani, S., Istaryatiningtias, and Qodariah, L. "A child-friendly school: How the school implements the model," *Int. J. Eval. Res. Educ.*, vol. 10, no. 1, pp. 273-284, 2021.

[14] Fauziati, E. "Child Friendly School: Principles and Practices," in *The First International Conference on Child - Friendly Education*, pp. 95-101.

[15] Anggraeni, C. D. and Nugraheni, T. "*Learning Saman Dance in the Formation of Students' Social Character*," vol. 419, pp. 7-10, 2020.

[16] Dini, T. A., Utomo, U. and & Triyanto, "The reflection of cultural identity of gayo society, central Aceh regency in tawar sedenge song," *Catharsis*, vol. 7, no. 2, pp. 189-198, 2018.

[17] Marhamah and Fauzi, "Islamic Dialectics and Local Culture in Petawaren Tradition of Gayo Community," *el Harakah*, vol. 2, no. 2, pp. 189-205, 2020.

[18] Indra, I., Rasyidin, A. and Sukiman, S. "Internalization of Falsafah 'Peri Mestike Gayo' as an Effort To Developing Students' Characters Based on Local Authority In SMA 1 Takengon Aceh Central District," *Int. J. Educ. Vocat. Stud.*, vol. 1, no. 8, p. 817, 2020.

[19] Syukri, S. "Budaya Sumang Dan Implementasinya Terhadap Restorasi Karakter Masyarakat Gayo Di Aceh," *MIQOT J. Ilmu-ilmu Keislam.*, vol. 41, no. 2, pp. 406-427, 2018.

[20] Hayati, E. "The Procession and Meaning of Sengen Tradition in Gayo Lues Society," *Adv. Soc. Sci. Educ. Humanit. Res.*, vol. 251, pp. 575-579, 2018.

[21] Anand, R., Singh, B., & Sindhwani, N. (2009). Speech Perception & Analysis of Fluent Digits' Strings using Level-By-Level Time Alignment. *International Journal of Information Technology and Knowledge Management*, 2(1), 65-68.

[22] Malik, S., Saroha, R., & Anand, R. (2012). A Simple Algorithm for reduction of Blocking Artifacts using SAWS Technique based on Fuzzy Logic. *International Journal Of Computational Engineering Research*, 2(4), 1097-1101.

[23] Juneja, A., Bajaj, S., Anand, R., & Sindhwani, N. (2020). Improvising Green Computing using Multi-Criterion Decision Making. *Journal of Advanced Research in Dynamical and Control Systems,* 12(3), Special Issue, 1161-1165.

[24] Kumar, M. S., Sankar, S., Nassa, V. K., Pandey, D., Pandey, B. K., & Enbeyle, W. (2021). Innovation and Creativity for Data Mining Using Computational Statistics. In *Methodologies and Applications of Computational Statistics for Machine Intelligence* (pp. 223-240). IGI Global.

[25] Pandey, D., Aswari, A., Taufiqurakman, M., Khalim, A., & Azahrah, F. F. (2021). System Of Education Changes Due to COVID-19 Pandemic. *Asian Journal of Advances in Research*, 10-15.

Chapter 3

Impact of COVID-19 on Various Sectors

Anjali Ahuja[1,*], Shikha Patheja[2,]
and Nidhi Sindhwani[3]
[1]Research, DCRUST, Murthal, India
[2]Research, PTU, Jalandhar, India
[3]AIIT, Amity University, Noida, India

Abstract

COVID-19 is a disease that belongs to the SARS COV-2 family and is highly infectious. Complete lockdown is the most common strategy employed by affected countries. Increased COVID-19 instances have affected people's health and daily lives and various sectors of the country. This chapter aims to explain the impact of COVID-19 on the various sectors. We examined ten various sectors in this study: financial institutions, agricultural and supply chains, tourism and hospitality, FMCG, Pharmaceuticals, Education, Railways, Delhi rail services, Power, and Telecom.

Keywords: COVID-19, hospitality, FMCG, education, Telecom

3.1. Introduction

Since the dawn of time, humanity has faced many obstacles, but the issues raised by COVID-19 are unique, and the consequences of these challenges have been significant. The good news is that most of humanity has been wiped

---

* Corresponding Author's Email: anjali.ahuja85@gmail.com.

In: An Interdisciplinary Approach in the Post-COVID-19 Pandemic Era
Editors: D. Pandey, B. Kumar Pandey, N. Sindhwani et al.
ISBN: 979-8-88697-052-4

out by this sickness. The COVID-19 situation has interrupted our way of life, but it's fair to say that this disruption has opened up many new prospects for us [1-4]. The current coronavirus (COVID-19) generated global concern in early January 2020, and by the end of March 2020, the outbreak had infected a large number of people all across the world [5]. On 30 January 2020, the first instances of COVID-19 in India were recorded in three locations in Kerala, among three Indian medical students who had returned from Wuhan, the pandemic's epicentre [6]. On 23 March 2020, Kerala declared a state of emergency, followed by the rest of the country on 25 March. India's recoveries surpassed active cases for the first time on 10 June. A second wave began in March 2021 and was far more damaging than the first, with vaccine shortages, hospital beds, oxygen cylinder shortages, and other medical supplies in some sections of the country.

India had surpassed estimates for new and active cases by late April. On 30 April 2021, it became the first country to record over 400,000 new cases in 24 hours [7]. Experts predict that the virus may spread to India instead of completely disappearing and becoming endemic [8].

India began its vaccination program with Covishield and Covaxin on 16 January 2021 [9, 10]. Later, the Sputnik V and Moderna vaccines were licensed for emergency use. As of 17 August 2021, the country had supplied approximately 550 million vaccine doses (Govt. of India). On 21 October 2021, at 9:47 a.m., India passed 100 crore doses, according to the Co-WIN website.

This paper addresses the key research question: what are the impacts of COVID-19 on various sectors? The remainder of the chapter structures are as follows: Section 2 represents the literature review. Section 3 defines the objective of the study. Section 4 describes the research methodology. Section 5 explains the impact of COVID-19 on various sectors. Section 6 defines conclusions and recommendations.

3.2. Review of Literature

Alexandra M. Stern et al. [11] explained virus was considered threatening for schools, movie halls, restaurants, and other places with heavy human traffic. Rules and orders were enforced in states and towns in various countries. Schools were closed in US, and people were asked to stay home for their health. This virus was observed as dangerous for children in US.

Madhav [12] explained that high mortality rates were associated with influenza in 1918. Various factors were involved, which increased the virus numbers and led to deaths. If a virus makes a comeback, it will be different, but elements may remain the same, which led to extinction due to the current COVID-19 viruses.

McKibbin and Fernando [13] used the CGE, i.e., a computable general equilibrium model. They found that overall GDP would be reduced by more than US $ 2 trillion in 2020 due to the pandemic, a case study about Hong Kong flu. There is the same flu in Spain and a fall in GDP of around the US $ 9 trillion. Due to the pandemic, the losses are severe all over the world.

Mishra [14] defined there is a threatening effect of COVID-19 in almost every sphere of the economy. It affects the industrial growth, international trade, employment, income of people, financial markets, businesses, and corporate offices. There is a massive loss in global trade, which affects economic growth. On the international front, there is a more than 30 percent loss. Hence there is enormous loss due to lockdown all over.

Singh et al. [15] explained the positive and negative impacts of the pandemic on the FMCG Sector. The changes that have come about are due to COVID-19. Production, large-scale adoption of cashless payment, etc., are described in this study. The authors define the models of CPG and FMCG. The study found that FMCG products would acquire the market quickly because they are moving with the help of the value chain.

Ozili and Arun [16] noted the severe effects of COVID-19. He stated that due to regulations regarding social distancing, there is a shutdown of the corporate world, financial markets, and many other businesses, which can help the country's economic growth. This closure of various markets led to heavy losses in different sectors of the economy and, ultimately, revenue reduction. There was an industrial decline in China of more than 13 percent during the starting phase of the pandemic.

Jaipuria et al. [17] examined how the pandemic has impacted not only foreign exchange earnings (FEE) but also regional advancements and job prospects, disturbing local communities. The investigation focused on foreign tourists' arrival in India and FEE using artificial neural networks (ANN).

The authors evaluated COVID-19's impact using four scenarios: lockdown and no lockdown in terms of FEE profit and loss.

Mittal and Sharma [18] examined the pandemic impact on health care and pharma stocks. This study has applied the widely used event study methodology to two-sector; calculated abnormal returns and cumulative abnormal returns and tested their significance. The authors revealed that there had been significant abnormal returns and cumulative abnormal returns in the healthcare and pharmaceutical sector.

3.3. Objective of the Study

The objective of this article is to study the impact of COVID -19 on the various sectors.

3.4. Research Methodology

We have collected data from online sources like government publications, journals, published reports, and newspapers. The search strings used to manage the data such as COVID-19, financial institutions, agriculture, FMCG, education, telecom, power, tourism, healthcare and rail services, etc.

3.5. Impact of COVID-19 on Various Sectors

3.5.1. Impact on Financial Institutions

COVID-19 has been identified as the century's black swan event, with far-reaching macroeconomic implications for India and the rest of the world [19]. The exponential spread of COVID-19 has caused a significant decline in major indexes, indicating its effect and potential to impact GDP growth significantly. While COVID-19 is expected to have an overall negative impact on credit growth in most sectors, the amount and nature of the impact will most likely vary depending on the length and breadth of the disruption. The BFSI industry in India received a boost in 2018-19, with the NPA ratio dropping more than 2% from 11% to around 9% as a result of improved recovery [19]. After increasing for ten consecutive years, this was a promising sign. Following the economic downturn caused by the COVID-19 issue, according to a recent assessment released in March by S&P Global Ratings, banks in the country are anticipated to see a 1.9 percent increase in their NPA ratio and a 130 basis point increase in credit cost ratios in 2020. On 27 March, the Reserve Bank of India (RBI) granted debtors of all finance institutions a three-month repayment moratorium on all term loans, credit cards, and overdrafts [20]. Under the established accounting rules, all lending institutions have the option to set policies for giving relief to eligible borrowers.

Moreover, there would also be an asset classification standstill for all accounts from 1 March to 31 May. This will reduce corporate pressure and result in no asset classification downgrade. RBI has also directed all the financial institutions to maintain a higher provision of 10% on all accounts under the standstill, spread over two quarters, i.e., March 2020 and June 2020. This announcement ensures that banks maintain sufficient buffers and remain adequately provisioned to meet future challenges. The Reserve Bank of India has also instructed all financial institutions to retain a more significant provision of 10% on all accounts subject to the standstill for two quarters, i.e., March 2020 and June 2020. This announcement is being made to ensure banks have enough buffers and provisioning to handle future problems. Figure 3.1 shows the NPA on scheduled commercial banks from 2011-2021.

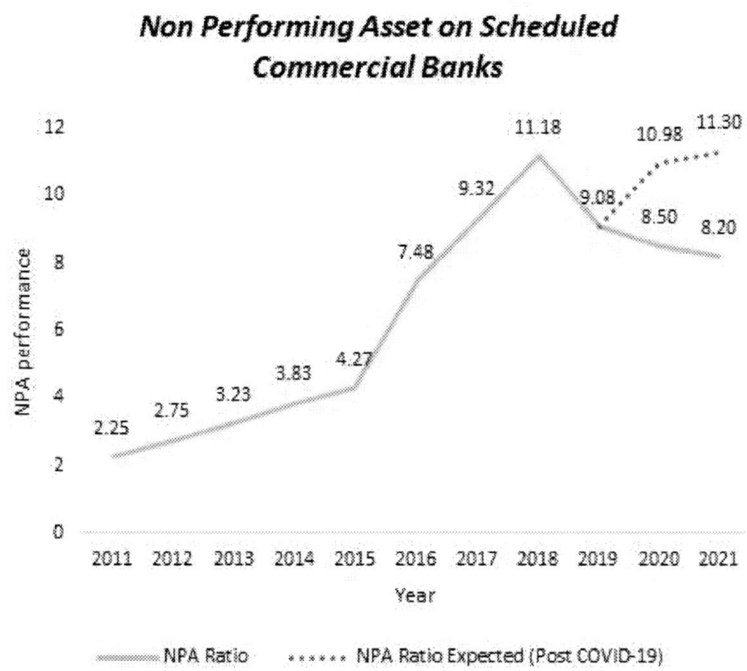

Figure 3.1. NPA on scheduled commercial banks.

3.5.2. Impact on Agricultural and Supply Chain

COVID-19 is causing some agricultural and supply chain disruptions [21, 22]. According to preliminary reports, the lack of migrant labour is causing specific harvesting processes to be disrupted, notably in northwest India, where wheat and pulses are picked. Supply chain operations are being disrupted as a result of transportation issues and other issues. Wheat, vegetables, and other crops have seen lower prices, but consumers still pay more. According to accounts in the media, the lockdown has already slowed milk sales by closing hotels, restaurants, confectionery shops, and tea shops. Meanwhile, disinformation, particularly on social media, has harmed poultry farms by implying that chickens are carriers of COVID-19.

3.5.3. Impact on Tourism and Hospitality

COVID-19, a pandemic proclaimed on 12 March, 2020 has significantly influenced the travel and hospitality industries [5]. As a result of the COVID-19 pandemic, the travel and tourism industry is estimated to lose 100.08 million jobs worldwide [23]. Tourism is the most dynamic sector, contributing to global economic growth and recovery by assisting many other sectors such as accommodation, catering, transportation, retail, entertainment, and so on [24]. Tourism is a significant source of revenue and employment in India. It is a source of employment, income, tax revenue, and foreign exchange earnings. The widespread COVID-19 has significantly impacted the tourism industry, which may continue for some years. According to Jaipuria et al. [17], the number of international tourists arriving in India from various regions decreased by 68 percent compared to the previous month. It significantly impacts tourism revenue in the form of foreign exchange fees.

3.5.4. Impact on the FMCG Sector

The FMCG industry is India's fourth-largest sector, accounting for half of all FMCG transactions in the country. Fast-moving Consumer Goods (FMCG) is the term used here. COVID-19's impending global outbreak has prompted the FMCG industry to declare an international emergency [25]. Following are the impact of COVID-19 on the FMCG sector:

- Prices to remain stable
- Shift towards local production
- Cashless to become a prominent trend
- Single brand and luxury retailers at high risk

3.5.5. Pharmaceutical Sector

According to McKinsey and Co. Report, 2020 India has the world's third-largest pharmaceutical sector by volume. In terms of revenue and employment, healthcare has become India's most important industry. Infectious diseases have an impact not only on persons but also on the economic and financial health of a nation [26-30]. The source of active pharmaceutical ingredients (APIs) is a critical aspect of the pharma industry's strategic plan to combat the COVID-19 pandemic, according to a report on the Indian pharmaceutical business. Most APIs used in generic medicine manufacture are obtained from India, which also supplies over 30% of generic APIs used in the United States. However, for the development of their medication formulations, Indian manufacturers rely primarily on APIs from China, with over 70% of their APIs coming from China, the world's largest producer and exporter of APIs by volume. The reliance of Indian pharmaceutical companies on Chinese APIs poses a severe threat to national health security, forcing the Indian government to establish a task force to examine the indigenous API sector. Several key representatives from the pharmaceutical industry and the NITI Aayog suggested that encouraging approvals of pharmaceutical infrastructure developments, clearance from the environment ministry, and tax exemptions and subsidies for developing and promoting pharmaceutical industry hubs could all help the market.

3.5.6. Impact on Education Sector

The lockdown has impacted various companies and workplaces all around the world. India, too, has been affected by this. There is a temporary shutdown, and employees are working from home. This shutdown has impacted students from different universities and schools. Students ranging from class 9 to 12 were most affected at the school level as these classes were considered the most crucial for students' careers. The board makes arrangements for online means for imparting education to students. Regularly scheduled classes were provided to students so they would not suffer any education loss due to lockdown [31]. This pandemic gas affected commercialized educational institutions due to their nature of business and revenue loss. However, institutions have started imparting education online via apps. Online courses were provided to students. Apps like Zoom, and Teams, were used to impart education to students. This pandemic affected teachers also as they were forced to learn the application of online apps to provide lectures to students. All teachers are not familiar with these apps, but teachers are trying to deliver their best. This has affected those students also who do not have a facility of mobiles or internet connections. Also, teachers are pressured to complete their courses before the examination. Many students were not attending classes online. There have been online examinations that have a significant impact on students' careers. The level of quality has been reduced through online education. Students and teachers using online mode may not have good internet connections, which reduces the quality of education. So education sector has been impacted a lot due to this pandemic.

3.5.7. Impact on Railway Sector

Due to this pandemic, there was a massive loss in the transportation sector. One of the most affected sectors is railways. The transportation industry has to control the movement of passengers further to reduce the spread of the virus [32]. As the railway is a high-density traffic mode, it is a great challenge for them to control the spread of the virus via rail traffic. On the other hand, the rail sector bears a considerable loss. There is an adverse effect of COVID-19 on railway operations. This loss occurs due to a shortfall in passengers and less loading. The revenue declined by 36 crores. Due to consequential lockdown, there were partial passenger services. This pandemic affects railways in the short term and long term. Once the pandemic is over, there will be a change in passenger behavior.

The rail department will have to look into new routines and technology for the future to increase passenger traffic. The train manufacturers need to be creative to survive in the future. New technologies should be introduced to limit the spread of the virus. Although the effect has hampered profits still railways, continue to operate with their available resources. Railways are facing challenges; however, around 12 lakhs employees work 24 hours to provide services to their customers. Due to heavy losses, this sector is looking forward to funding sources, including private investments [32].

3.5.8. Impact on Metro Rail Services

Delhi Metro suffers a massive loss due to COVID 19. More than 16 crores loss was suffered by Delhi Metro due to the closure of services during pandemic times. Services were also suspended all over the country. During the lockdown, when services were not there, they prepared for their tender schedules, finalized designs, and finalized tenders. To mitigate the losses incurred, the Delhi metro rail network will enhance revenue through various innovative means like providing the facility of a feeder system to passengers, giving space for lease, property development at stations, and enhancing finance [33]. The Delhi metro tries to ensure financial stability during metro rail operations. Metro rail services include Delhi metro, Bengaluru Metro, Hyderabad Metro, Kochi Metro, Ahmedabad Metro, Chennai Metro, Mumbai Metro, Kolkata Metro, Lucknow Metro, and Jaipur Metro.

3.5.9. Impact on Power Sector

The power sector is one of the critical sectors for productivity. It has weakened the global economy after a sharp fall in demand due to the lockdown. The lockdown has already impacted the power sector in the short term. There was a complete shutdown all over except the essential commercial activities. The electricity demand from commercial and industrial customers decreased significantly while the residential demand increased [34]. The electricity regulatory authorities ensure to distribute power at lower rates for domestic and agricultural consumers, which is very low compared to industrial consumers.

This lockdown has impacted the coal-based plants. According to reports, the thermal plants bear losses and decline by 40 percent. This lockdown poses a severe challenge to the power sector worldwide.

The power sector has been struggling with its finance, as the losses were more and financial management is complex. The power sector delayed the payment. The challenges before this sector were threatening as they could push it over the edge. The power sector seeks funding support to reduce the losses borne by them.

3.5.10. Impact on Telecom Sector

The Telecom sector is a fast-growing sector in India. An increase in COVID-19 led to a lockdown and imposed restrictions on movement. Companies started working from home for their employees, which further led to more data usage for work, which has a significant impact on the Telecom sector. The telecommunication companies are focussing on increasing networks and reliability for their customers. COVID-19 has impacted their plans for 5G networks [35]. However, technologies like remote working, video conferencing, and telecommunications have increased as employees work from home. Entertainment sources have also emerged like Netflix, amazon prime, and Hotstar. This all has put the spotlight on the telecom sector. The telecom sector has helped the government in many contemporary issues. The government has taken help from the telecom sector in spreading the awareness regarding COVID-19 among people. The telecom service providers help track the location of COVID- 19 patients. More internet usage has decreased the speed of mobile and broadband in India [36] as this facility has more customer traffic due to working from home. Telecommunication has also given notice for setting up more towers. The telecom sector must develop its infrastructure and sound quality tools for its customers. COVID-19 has increased the dependency of customers on digital devices.

Conclusion and Recommendation

The impact of COVID- 19 is incredible and will disrupt the economy until the world gets vaccinated. This pandemic resulted in a massive loss of millions of jobs in all sectors of the economy. It has severe impact than any other pandemic. The most affected industries are tourism, oil, FMCG, education, railways, and many more. Unemployment is more at this time and is considered the worst recession ever. COVID-19 is a nightmare for human beings worldwide as this pandemic results in more death rates [37]. The vaccination has come, and people are getting vaccinated.

These sectors must take various measures so that economy will grow. One of the essential measures to boost the economy again is innovative technologies that help customers attract these sectors again [38, 39]. Business units must optimize the workforce. These sectors require financial support so that their businesses can revive and help in economic growth.

One of the measures that these sectors can take is digitization. Digital technologies can help these business units to get more revenue. This pandemic has a considerable effect, but by introducing innovative techniques, industries can generate income.

References

[1] Rather RA, Islam T, Rehman IU and Pandey D. (2021). Development of vaccine against coronavirus disease 2019 (COVID-19) In India. *Asian Journal of Advances in Medical Science*, 13-21.

[2] Meivel S, Sindhwani N, Anand R, Pandey D, Alnuaim AA, Altheneyan AS, Jabarulla MY and Lelisho ME. (2022). Mask Detection and Social Distance Identification Using Internet of Things and Faster R-CNN Algorithm. *Computational Intelligence and Neuroscience*, 2022.

[3] Jain S, Sindhwani N, Anand R and Kannan R. (2022). COVID Detection Using Chest X-Ray and Transfer Learning. *International Conference on Intelligent Systems Design and Applications*, (pp. 933-943). Springer, Cham.

[4] Kadam SA and Pandey D. (2020). Post COVID-19 Opportunities and Challenges Formicro Small Medium Enterprises (MSMEs) Development in India. *Asian Journal of Advances in Research*, 1-4.

[5] WHO. (2020). *Coronavirus disease (COVID-19) pandemic.* https:// Coronavirus disease (COVID-19) (who.int).

[6] Andrews MA, Areekal B, Rajesh KR, Krishnan J, Suryakala R, Krishnan B and Santhosh PV. (2020). First confirmed case of COVID-19 infection in India: A case report. *The Indian Journal of Medical Research*, 151(5), 490.

[7] Coronavirus | India becomes first country in the world to report over 4 lakh new cases on 30 April 2021. *The Hindu*. Special Correspondent. 30 April 2021. ISSN 0971-751X. Retrieved 2 May 2021.

[8] India's first coronavirus patient discharged after being cured". *Hindustan Times.* 20 February 2020. Retrieved 24 July 2021.

[9] COVID vaccine doses administered in India cross 55 crores: Govt. Business Standard India. *Press Trust of India*. 16 August 2021. Retrieved 17 August 2021.

[10] Pandey D, Islam T and Malik MA. (2021). Novel coronavirus disease (SARS-COV-2): An overview. *Asian Journal of Advances in Medical Science,* 8-12.

[11] Stern AM, Cetron MS, Markel H. (2009). *Closing The Schools: Lessons From The 1918-19 U.S. Influenza Pandemic*, http://hlthaff.28.com.

[12] Madhav N. (2013). Pandemics: Risks, Impacts, and Mitigation. *Journal of Research,* 12(8), 999-1020.

[13] McKibbin WJ, Roshen F. (2020). The Global Macroeconomic Impacts of COVID-19: Seven Scenarios. *Journal of International Development*, 11(2), 22-33.

[14] Mishra A, Patnaik P. (2020). *COVID-19 pandemic*! It's impact on people, economy, and environment, Wiley online library.
[15] Singh VK, Pawar DS, Shekam L and Dutt V. (2020). Impact of COVID 19 on FMCG Sector. *Journal of Critical Reviews*, 7(12), 4477-4484.
[16] Peterson KO and Thankom A. (2020). Spillover of COVID-19: Impact on the Global Economy. *Social Science Research network,* https://ssrn.com/abstract=3562570.
[17] Jaipuria S, Parida R and Ray P. (2021). The impact of COVID-19 on tourism sector in India. *Tourism Recreation Research*, 46(2), 245-260.
[18] Mittal S and Sharma D. (2021). The impact of COVID-19 on stock returns of the Indian healthcare and pharmaceutical sector. *Australasian Accounting, Business and Finance Journal*, 15(1), 5-21.
[19] Michie J. (2020). The covid-19 crisis–and the future of the economy and economics. *International Review of Applied Economics*, 34(3), 301-303.
[20] Patnaik I and Sengupta R. (2020). Impact of COVID-19 on the Indian economy: An analysis of fiscal scenarios. *Indian Public Policy Review*, 1(1 (Sep-Oct)), 41-52.
[21] Shahed KS, Azeem A, Ali SM and Moktadir M. (2021). A supply chain disruption risk mitigation model to manage COVID-19 pandemic risk. *Environmental Science and Pollution Research*, 1-16.
[22] Shukla R, Dubey G, Malik P, Sindhwani N, Anand R, Dahiya A and Yadav V. (2021). Detecting crop health using machine learning techniques in smart agriculture system. *Journal of Scientific and Industrial Research (JSIR),* 80(08), 699-706.
[23] Saini R. (2020). *Impact of Corona Virus on Indian Economy*. Available at SSRN 3595300.
[24] Liu H, Liu Y, Wang Y and Pan C. (2019). Hot topics and emerging trends in tourism forecasting research: A scientometric review. *Tourism Economics,* 25(3), 448-468.
[25] Rajamohan S, Jenefer J and Sathish A. (2021). Impact of COVID-19 on FMCG Sector. *Shanlax International Journal of Management*, 8(4), 69-74.
[26] Mayer JD. (2000). Geography, ecology and emerging infectious diseases, *Social Science and Medicine*, 50(7): 937-952. http://dx.doi.org/10.1016/S0277-9536(99)00346-9.
[27] Zhang Y, Dang Y, Chen H, Thurmond M and Larson C. (2009). Automatic online news monitoring and classification for syndromic surveillance. *Decision Support Systems,* 47(4): 508-517. http://dx.doi.org/10.1016/j.dss.2009.04.016.
[28] Anand R, Sindhwani N and Saini A. (2021). Emerging Technologies for COVID-19. Enabling Healthcare 4.0 for Pandemics: A Roadmap Using AI, *Machine Learning, IoT and Cognitive Technologies*, 163-188.
[29] Gupta A, Anand R, Pandey D, Sindhwani N, Wairya S, Pandey BK and Sharma M. (2021). Prediction of Breast Cancer Using Extremely Randomized Clustering Forests (ERCF) Technique: Prediction of Breast Cancer. *International Journal of Distributed Systems and Technologies (IJDST),* 12(4), 1-15.
[30] Khongsai L, Anal TS, AS R, Shah M and Pandey D. (2021). Combating the spread of COVID-19 through community participation. *Global Social Welfare,* 8(2), 127-132.

[31] Pandey D, Ogunmola GA, Enbeyle W, Abdullahi M, Pandey BK and Pramanik S. (2021). COVID-19: A framework for effective delivering of online classes during lockdown. *Human Arenas*, 1-15.

[32] Tardivo A, Carrillo Zanuy A and Sánchez Martín C. (2021). COVID-19 impact on transport: A paper from the railways' systems research perspective. *Transportation Research Record*, 2675(5), 367-378.

[33] Tayal D and Mehta AK. (2021). Working women, Delhi Metro and COVID-19: a case study in Delhi-NCR. *The Indian Journal of Labour Economics*, 64(2), 389-413.

[34] Saini MK, Nagal R, Tripathi S, Sindhwani N and Rudra A. (2008). PC Interfaced Wireless Robotic Moving Arm. In *AICTE Sponsored National Seminar on Emerging Trends in Software Engineering*, (Vol. 50).

[35] Kirubasri G, Sankar S, Pandey D, Pandey BK, Singh H and Anand R. (2021, September). A Recent Survey on 6G Vehicular Technology, Applications and Challenges. In *2021 9th International Conference on Reliability, Infocom Technologies and Optimization*, (Trends and Future Directions) (ICRITO) (pp. 1-5). IEEE.

[36] Anand R, Shrivastava G, Gupta S, Peng SL and Sindhwani N. (2018). Audio watermarking with reduced number of random samples. In *Handbook of Research on Network Forensics and Analysis Techniques*, (pp. 372-394). IGI Global.

[37] Ayenew B, Pandey D, Yitayew M, Etana D, Binay Kumar P and Verma N. (2020). Risk for surge maternal mortality and morbidity during the ongoing coronavirus pandemic. *Med Life Clin*, 2(1), 1012.

[38] Singh H, Rehman TB, Gangadhar C, Anand R, Sindhwani N and Babu M. (2021). Accuracy detection of coronary artery disease using machine learning algorithms. *Applied Nanoscience*, 1-7.

[39] Sindhwani N, Verma S, Bajaj T and Anand R. (2021). Comparative Analysis of Intelligent Driving and Safety Assistance Systems Using YOLO and SSD Model of Deep Learning. *International Journal of Information System Modeling and Design (IJISMD)*, 12(1), 131-146.

Chapter 4

Consequences of the COVID-19 Pandemic on Web-Based Educational Processes: Benefits, Problems and Suggestions

Rahul Kumar Nandi*
Department of Education, Fakir Mohan University,
Balasore, Odisha, India

Abstract

With the rising of COVID-19 pandemic situation, the educative process of the whole world undergoes a drastic change. The lockdown process evolves to fight against the COVID-19 pandemic which creates tremendous change in education sector. The teaching institutions are bound to close their physical mode of teaching instructions and shift towards web-based educational process. The traditional chalk and blackboard classrooms transform into technology-based e-learning classrooms. This stops the individual one-to-one live interaction of students and teacher and move to distant type of learning process with the introduction of various digital delivery model. This new approach of instruction delivery ensured uninterrupted educational delivery in the schools as well as in higher educational institutions. This review discusses the various issues faced by the teaching institutions during online learning and assessment process in this COVID-19 pandemic. It also provides some suggestive measures for improvement and how to deal withthe problematic situations of web-based learning and assessment system. This study also summarizes the beneficial aspects of e-learning which helps in reshaping the future aspects of education.

---

* Corresponding Author's Email: tapas.nandi174@gmail.com.

In: An Interdisciplinary Approach in the Post-COVID-19 Pandemic Era
Editors: D. Pandey, B. Kumar Pandey, N. Sindhwani et al.
ISBN: 979-8-88697-052-4

Keywords: COVID-19 pandemic, web-based learning, assessment, teaching institution, educational process, e-learning

1. Introduction

With the passage of time every system requires some sort of changes. This is found to be seen in the current era of COVID-19 pandemic [1-5] which greatly forced to reshape the process of education. For more than one year, the entire globe had been confrontedan exceptionallockdown situation as a consequence of first and second waves of pandemic COVID-19. All the schools have closed to avoid the infection of the students and staff almost. Because all schools are closed indefinitely, teachers are experimenting with various methods to reduce the effect of COVID-19 on the educational process. Education is the backbone of any nation, and stopping educational activities can adversely affect the development of the nation [6, 7]. Therefore, the educational administration tried to continue the education by moving from traditional teaching-learning to an online platform. Consequently, to continue the education, the institutions across the globe, from schools to universities, moved virtual mode/online mode [8-10]. However, the significant concerns for teaching and learning quality are closely related to how well the content is designed and executed [11]. Besides this, the effectiveness of online teaching is directly related to the understanding of constraints faced by teachers and students and curation of content for the online environment. However, many studies show that the online teaching-learning process faces various issues, challenges, and constraints during this COVID-19 pandemic [12-14]. At the same time, the school teachers are struggling to function in a COVID-19 pandemic as effectively as in a normal situation in the absence of an actual classroom setting through online mode. The issues and challenges faced by school teachers are much broader in comparison to the higher institution teachers [15, 16]. In this line, we have examined high school teachers' experiences regarding the web-based teaching methods during the COVID-19 pandemic and their suggestions to make it more effective and successful.

2. Literature Review

The various teaching institutions and organizations apply e-learning in their teaching approach but they first trained teachers with capability building training platform. The web-based learning skills provide the delivery of instruction at a reduced amount than physical modes of classroom teaching. Thus, it lay out more scopes for students to admitting into educational programs [17]. Many types of research were conducted to examine the behavior of teachers and students toward the online. A sudden surge of COVID-19 pandemic impetus teaching institutions to implement e-learning without proper planning and preparation. At the initial stage, most of the teachers and other stakeholders were neither prepared to face this unprecedented situation of the COVID-19 pandemic nor skilled to use the online platform to teach students virtually. However, various educational institutions (Universities, colleges, and other educational organizations) organized workshops, faculty development programs, short-term courses, and webinars to train the teachers to teach virtually. These courses played a vital role in enabling the teachers to effectively and efficiently utilize the available online resources in their teaching-learning process [14]. The web-based learning become more beneficial and suitable for the learning society, if necessary, steps must be taken in the field of arrangement of digital platform and resources by the administrators as well as educational institutions. The more use of ICT resources helps the students to enhance the skills in the field of technology which is a prime requisite in the present days for all [18]. In order to revolutionize the existing education system, the educational institutions runed by both government and private agencies should organizing proper training systems for all levels of education before implementing the web-based learning. This will help to reduce the stress level of teachers as they have to face the new approaches of teaching with a sudden change [19]. After adjusting with the technological based learning system, it seems a positive and developmental aspects in the field of education. So, the curriculum planner must integrate technology to the classrooms during post COVID approach also. This provides the way in extending education process to a new level harmonizing with the global context. As a result, both educators and learners developed a favorable perfection towards e-learning and not face much problem while adopting it [20]. Various studies revealed that the educators designed their content using ICT resources for delivering lessons and assessing the students. They utilize the web-based teaching and assessment techniques in order to maintain equilibrium with the modernized context of education [21]. The Google classroom, Google meet, zoom app, WhatsApp, YouTube, etc., become essential tools in the hand of teachers to mitigate the distress of the online teaching-learning process. These technological teaching-learning tools enabled educational institutions to pursue creative approaches in education in the concise notice. The educational institutions are asserting to utilize these technological tools to continue the educational system, but it is important to note that the learning quality depends on digital access and efficiency of both students and teachers [11].

3. Research Methodology

The prevailing two waves of the coronavirus pandemic situation have manipulated the traditional learning style with the current web-based methods of learning. The present research study is based on the analysis of a literature survey focusing on the digitalized education process [22] during the existing COVID-19 period. This review paper is formulated on the basis of recent findings and publications related to the major emphasis on benefits, problems, and issues in online methods of teaching and assessment procedures.

4. Analysis and Discussion

4.1. Effect of COVID-19 Pandemic on Educational Process

The COVID-19 pandemic has generated the most significant disturbance in the whole educational system, affecting nearly 1.6 billion learners in more than 190 countries and all continents [23]. Nearly all schools have been closed for the last one and a half years due to the unprecedented situation of human history. Every country framed some new policies to cope with the effect of the COVID-19 pandemic on teaching institutions by using ICT resources, especially online resources. At the same time, the country faced a complete lockdown for many months and partially to date. The educational institutions, including schools, are closed, and most schools use ICT resources to continue the instructional teaching methos. However, the ability to respond to school closure is proportional to the school's level. The good resources schools are using ICT resources entirely to dominate education among their students, while the resources crunched schools, especially in rural areas, face problems to cope with the problems of the COVID-19 pandemic. This interruption in the teaching institutions will disproportionately negatively impacton the learners. They are central part of every educational system and the learning place for them at home is limited so it is essential to take initiatives to continue their education. The teachers and students use various ICT resources, i.e., social networking, WhatsApp group, Google Meet, Google Classroom, Moodle, Zooms, you tube, etc., in the process of education.

4.2. Benefits of Web-Based Teaching and Learning

Online learning provides several benefits to students, including position flexibility and well-managed time, enrichment of teaching resources, elimination of barriers that hinder student participation, use of discussion forums, motivates students for alternative instructional approaches, cost-effectiveness, consideration of individual differences, and learning at their own pace and time [26]. The authors in [27] described the advantages of online learning according to the learners' points of view; accessibility, protection of time, and decreased price are welcomed as major helpful properties that drive them towards implementation and prolonged use of web-based learning. Online learning also supports advanced mental ability students, nurturing self-reflection and abstract thinking and levelling the playing field between participants.

The COVID-19 pandemic forced schools to move from traditional to an online teaching-learning environment. The authors in [28] mentioned various benefits of web-based teaching during the COVID-19 pandemic, some of which are stated below:

- Students obtain education while residing at home and are also able to maintain social distance.
- Socialdistance reduces the chance and possibility of COVID-19 infection.
- Children get the scope to learn in suitable learning environments.
- It provides complimentary collaborative provision, which permits learners to study at any speed suitable to them.
- The scheduling of teaching is not a rigid one.

- It offers education to students at any place and at any time by using their electronic devices (mobiles, computers, tablets, etc.) to connect with the internet.
- Though it is online teaching methods but educators monitor and report students' progression by providing effective feedback just like to physical mode.
- It expands the role of teaching institutions by utilizing technology for solvingactual difficulties facing in real life.

The COVID warriors like police, doctors, and nurses, who does not get opportunity to take part in e-learning teaching utilize class video recording of classes to continue their studies throughout the pandemic. Figure 1 shows the benefits of web-based teaching and learning during COVID-19.

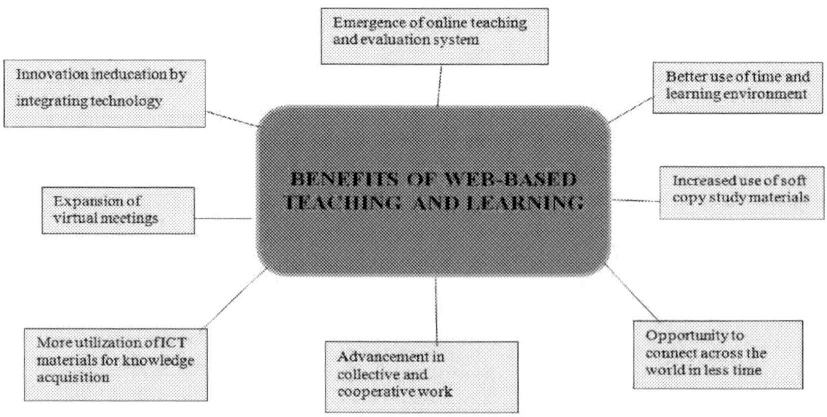

Figure 1. Benefits of web-based teaching and learning during COVID-19.

The support and collaboration of all stakeholders, i.e., government, policy makers, curriculum developers, school administration, experts, teachers, parents, community, and students are necessary to benefit from an online teaching-learning environment [29].

Despite these advantages, online teaching-learning has several disadvantages, such as the learners feel isolated due to lack of interactions, a distraction from the path of learning due to the availability of tons of distracting content, and lack of hands-on activities. The ability of teachers and students to use the technology effectively in teaching and learning is another significant limitation of online teaching [27]. These drawbacks are quite obvious for school-going children who may not get the scope to use ICT instruments like computers and smart phones and also not acquaintedwith the e-learning methods and instructions.

These limitations are more evident for school-going students who may not have online access or have limited experience with online learning tools, such as computers. The lack of needed skills for online teaching-learning can make online teaching-learning less effective than the traditional teaching-learning method.

4.3. Major Problems of Online Teaching and Learning Process

Every new thing while gaining entry to the current system meets a variety of problems and difficulties. To remain completely existent requires a little bit of time. Similarly, the teaching institutions along with their all-stake holders find themselves confronted while using a web-based learning approach. There are various issues that a teacher and students can face during the online teaching-learning process. Some of them are discussed here.

1. *Network, battery, and power supply issues*: the most significant issues teachers and students face with ICT devices. Sometimes the network does not work smoothly and causes disturbances in communication between teacher and student. The teacher often feels that students are not listening to him and just talking to the screen. Besides this, the mobile battery and power supply can also hamper the teaching-learning process. Frequent power outages can impede mobile charging and put a halt to classroom instruction.
2. *Accessibility of ICT devices*: This is another big issue that is hindering the teaching-learning process online. Teachers of significant and sound schools would have these ICT devices, but rural and small-town teachers do not have sufficient resources to purchase these devices. In the same way, some families cannot afford smartphones or other digital devices due to their financial conditions. This issue is more severe with the students of government schools because most students in government schools belong to the economically weaker section (EWS) of society. Most of the families belonging to EWS generally have a single mobile phone or simple feature phone owned by their father, and at a time, only one person can use it. It became a challenge for a teacher to organise the online class on time. Many parents bring their phones with them and return home at night,

causing many students to miss class and fail to complete their homework on time.

3. *Use of teaching-learning applications on mobile and other digital devices*: This is another issue with online learning. Almost all the teachers initially faced problems using digital devices and tools in the process of educational instruction, but they gradually learnt how to use these devices in the teaching-learning process. However, this is the accepted view that every ICT device and tool has different functions from others; they are not fully skilled in using those functions. They face problems presenting the screen, preparing the PowerPoint presentations, preparing videos for students, creating Google forms, and sharing links with students. These things are new to them. That is why some of the teachers give lectures like they are reading in front of the screen. These problems are not only faced by the teachers but also the students. Sometimes the students are unable to understand what to do with ICT tools. They had difficulties in joining the class; they did not know how to keep the mic off or on and when to respond in onlineclass.
4. *Time management at home and privacy*: Using ICT tools, especially online classes from home, creates many time management problems for the teachers. At home, they have some other responsibilities and family members, such as small kids. Sometimes these things create a hurdle for them to take or continue online classes. In the same way, the home environment of students can also create disturbances in their learning process. The teacher's privacy, predominantly female, is also at higher risk. As the teacher has to share their personal contact with the students in a WhatsApp group, the students can misuse it and make fun of the teachers.
5. *Issues related to assessing students' progress*: The process of teaching-learning is incomplete without assessing the students' progress. The online teaching-learning process is entirely different from the traditional teaching-learning process. Teachers use different assessment approaches to assess the students' progress, i.e., sending homework on WhatsApp, viva voce, providing assignments and projects. These are some approaches that teachers are using now. Students and teachers both face the problem in the case of of assessment. The teacher fails to measure the exact learning outcomes that the students have achieved in a specific course content.

6. *Health issues*: During this pandemic, the use of screens has increased manifold for teachers and students. This increase in screen time causes mental health problems for the teachers as well as the students [30-32]. Due to increased screen time, teachers and students may experience headaches, eye pains, posture deformity, stress, and other symptoms.However, teaching online, providing learning materials to students, and even responding to their queries at text affects the teacher's whole routine and health, such as strain in the eyes, headaches, and backpain.
7. *Students' discipline*: The COVID-19 pandemic made students careless, as students now do not take the class seriously. If they have to submit an assignment, they make excuses like network issues, or we do not have rough pages, or we can not go outside due to Corona, or we do not or did not have power in the phone, etc. In some cases, it can be a genuine reason, but in several cases, it is a fake excuse. Sometimes, students think the teacher cannot use the online teaching-learning platform properly to make fun of them. Sometimes they ruin the whole class with fun, like this:

The teacher is teaching online, and a student started shouting, "Hello, hello, hello... anyone listening to me? Is anyone listening to me?"

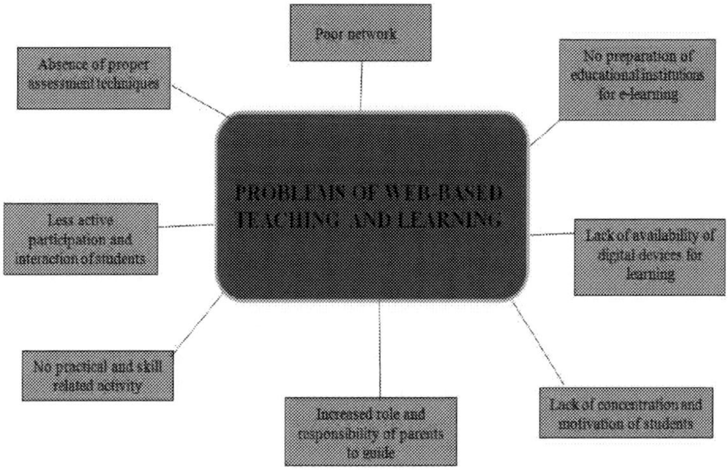

Figure 2. Problems of web-based teaching and learning during COVID-19.

Sometimes, this is most irritating for a teacher because one student ruins his/her class discipline even though he/she can hear everything and it is known by other students also. Figure 2 demonstrates the problems of web-based teaching and learning during COVID-19.

4.4. Suggestions for Improvement in the Web-Based Teaching-Learning Process

To become more efficient, the education system has to accept the recommendations in a positive way. One can learn and develop more by analysing the previous experiences gained during the challenge period and integrating the suggestions. Thus, all the educators are provided some suggestions to make the online systems of teaching more efficient during this pandemic period of COVID-19. Those suggestions can be summed up as follows:

- Teachers have to motivate students from time to time to participate actively in the online classroom.
- Teachers should take care of time in online classes as students sit in front of a screen or mobile phones for so long, so they should try to keep classes short.
- To resolve the student's carelessness during the online classes, teachers should inform students' parents to check their child's activities from time to time.
- Teachers should encourage the students to write in the chat box about their problems if they have any. This will not derail the whole online class from its original path.
- Teachers must set some rules and regulations similar to their offline class, such as asking students to mute and unmute whenever asked.
- To overcome students' absenteeism online, the teacher should ask students randomly to pen their video during online teaching.
- Teachers should create attractive teaching-learning materials like videos, PowerPoint presentations, etc. The videos should be short and simple. Teachers also search various online sources, such as YouTube videos, and provide those links to the students. In creating these learning materials, the teacher must consider the following: Why do

  - we use this as a teaching-learningmaterial?
  - What other resources can they use?
  - How to include such teaching materials into their teaching learning process is not clear.
- Teachers should change their pedagogical approach and include powerpoint presentations, quizzes, Google forms, animations, simulations, etc.

4.5. Suggestions for Evaluation System Measures and Strategies

Evaluation plays a very significant role in the teaching-learning process. Without assessment and evaluation, the educational process is incomplete. So, to know the real achievement of students in their respective fields of education, an educator has to correctly evaluate the learning outcomes. In this pandemic period, teachers are not able to judge properly the performance of learners. So, the educators must facilitate students and provide some strategy by which they can evaluate themselves. This provides the basis for students to be more competent and more skillful in the field of education. Ultimately, it will help to create motivation and interest among the learner to attain more knowledge and skills. Feedback is one of the most important attributes for the child's development, which is also not correctly provided after each assessment process. The regular and effective feedback process is also stopped due to this COVID-19 pandemic situation. Thus, the students are not able to track their learning progress and the teacher cannot detect the effectiveness of their methods and strategies adopted in the instructional process. Some cases have also been discovered in which, on the recommendation of both the national and state governments, there must be promotion to a higher class without any assessment system. Students are not properly screened for promotion, and their flaws and problems remain with them.Some cases may arise in these types of situations where students are unable to determine the criteria on which they are evaluated and certified. So, to make the evaluation process more effective and beneficial, the evaluators must use different strategies to assess and evaluate the student's progress. The teachers can use the following strategies:

- Provide scenario-based learning to students.
- Avoid direct questions and scenario-based assessment.

- Use of concept and understanding based questions.
- Ask them to prepare short videos on some topic.
- Use objective questions rather than subjective ones.
- Critical thinking skills are required to answer questions.
- Use quizzes to assess the students' progress.
- Ensure that the students do not copy and paste in writing assignments.
- to prepare some creative project-based work.
- Instruct to participate in discussion as a part of the assessment criteria.
- Provide learning materials and ask them to write a summary of them.
- Alternative methods of feedback must be used.

If we use various strategies to assess the students' progress, we can engage them actively in the teaching-learning process. Active involvement and active participation of students must be required for the effectiveness of teaching. Then the educational systems proceed in an advanced way and the learners provide a positive contribution towards the development of the nation.

Conclusion

The teachers' reflections revealed that they are facing various issues in the online teaching-learning process. However, this new context of teaching-learning is accurate, and it enables the teaching-learning process to be continued in this pandemic era when other things were closed entirely. Everything in nature has some positive as well as negative aspects. Similar situation also happened in the case of COVID-19, it has both valuable and deleterious effect. The reshaping in the educational instructions and procedures provide an indication to balance with the modern technological based education system. The modern trends of teaching and assessment techniques are developed and integrated to the old process of learning. This up-to-date e-learning system indeed advantageous emphasizing specifically for the developmental of curricular aspects of education. The knowledge gained through digital technology encapsulated that it discovers the recent techniques, methods and approaches of teaching as well as evaluation. But the web-based learning system gives less attention towards the co-curricular part like arts, craft and physical education, etc. So, the students are lacking skills related to practical and utility approach. The children of rural and remote areas find lots of complexities for using e-learning due to unavailable of ICT devices and low internet connection [33-35]. There is a need for more systematic, shared, and strategic approaches to harness the potentials of the online teaching-learning process. In place of focusing on negative aspects of the COVID-19, our focus should be to strengthen the partnerships among different stockholders of the education. Thus, the we-based learning methodology while incorporate with the conventional methods of education it become more successful. The potentials of this partnership can mitigate the adverse effects. In this way, the teachers and other stockholders of the education can enhance the quality of instruction for their learners by providing support, guidance, and a helpful hand.

References

[1] Jain, S., Sindhwani, N., Anand, R., & Kannan, R. (2022). COVID Detection Using Chest X-Ray and Transfer Learning. In: *International Conference on Intelligent Systems Design and Applications* (pp. 933-943). Springer, Cham.

[2] Ayenew, B., Pandey, D., Yitayew, M., Etana, D., Binay Kumar, P., & Verma, N. (2020). Risk for surge maternal mortality and morbidity during the ongoing corona virus pandemic. *Med. Life Clin.*, 2(1), 1012.

[3] Meivel, S., Sindhwani, N., Anand, R., Pandey, D., Alnuaim, A. A., Altheneyan, A. S., & Lelisho, M. E. (2022). Mask Detection and Social Distance Identification Using Internet of Things and Faster R-CNN Algorithm. *Computational Intelligence and Neuroscience*, 2022.

[4] Khongsai, L., Anal, T. S., AS, R., Shah, M., & Pandey, D. (2021). Combating the spread of COVID-19 through community participation. *Global Social Welfare*, 8(2), 127-132.

[5] Anand, R., Sindhwani, N., & Saini, A. (2021). Emerging Technologies for COVID-19. Enabling Healthcare 4.0 for Pandemics: A Roadmap Using AI, Machine Learning, IoT and Cognitive Technologies, 163-188.

[6] Idris, F., Hassan, Z., Ya'acob, A., Gill, S. K., & Awal, N. A. M. (2012). The role of education in shaping youth's national identity. *Procedia-Social and Behavioral Sciences*, 59, 443-450.

[7] Bansal, R. (2020). Is education the most important step in the development of the country? Accessed on 21.06.2021 from https://timesofindia.indiatimes.com/blogs/voices/is-education-the-most-important-step-in-the-development-of-the-country/.

[8] Todd, R. W. (2020). Teachers' perceptions of the shift from the classroom to online teaching. *International Journal of TESOL Studies*, 2(2), 4-16.

[9] Ray, S., & Srivastava, S. (2020). Virtualization of science education: a lesson from the COVID-19 pandemic. *Journal of Proteins and Proteomics*, 11, 77-80.

[10] Verma, G., Campbell, T., Melville, W., & Park, B. Y. (2020). Science teacher education in the times of the COVID-19 pandemic.

[11] Muthuprasad, T., Aiswarya, S., Aditya, K. S., & Jha, G. K. (2021). Students' perception and preference for online education in India during COVID-19 pandemic. *Social Sciences & Humanities Open*, 3(1), 100101.

[12] Allen, J., Rowan, L., & Singh, P. (2020). Teaching and teacher education in the time of COVID-19. *Jornal of Teacher Edcation.*

[13] Brank, E. M., & Scott, L. (2014). Parental responsibility for juveniles' acts. *The Encyclopedia of Criminology and Criminal Justice,* 1-5.

[14] Kulal, A., & Nayak, A. (2020). A study on perception of teachers and students toward online classes in Dakshina Kannada and Udupi District. *Asian Association of Open Universities Journal.*

[15] Parthiban, K., Pandey, D., & Pandey, B. K. (2021). Impact of SARS-CoV-2 in online education, predicting and contrasting mental stress of young students: a machine learning approach. *Augmented Human Research*, 6(1), 1-7.

[16] Zargar, S. A., Islam, T., Rehman, I. U., & Pandey, D. (2021). Use of Cluster Analysis To Monitor Novel Corona Virus (COVID-19) Infections In India. *Asian Journal of Advances in Medical Science*, 1-7.

[17] Murday, K., Ushida, E., & Ann Chenoweth, N. (2008). Learners' and teachers' perspectives on language online. *Computer Assisted Language Learning,* 21(2), 125-142.

[18] Gautam Kumar, *International Journal of Advanced Trends in Computer Science and Engineering*, 9(5), September-October 2020, 7704-7711.

[19] Garg, Sunayana & Diwakar, Aggarwal & Upadhyay, Sushil & Kumar, Gautam & Singh, Gulbir. (2020). Effect of COVID-19 on school education system: challenges and opportunities to adopt online teaching and learning. *Humanities & Social Sciences Reviews.* 8. 10-17. 10.18510/hssr.2020.862.

[20] Khan, M. A., Vivek, Nabi, M. K., Khojah, M., Tahir, M. Students' Perception towards E-Learning during COVID-19 Pandemic in India: An Empirical Study. *Sustainability* 2021, 13, 57. https://dx.doi.org/10.3390/su13010057.

[21] Bordoloi, R., Das, P., & Das, K. (2021). Perception towards online/blended learning at the time of COVID-19 pandemic: an academic analytics in the Indian context. *Asian Association of Open Universities Journal.*

[22] Kirubasri, G., Sankar, S., Pandey, D., Pandey, B. K., Singh, H., & Anand, R. (2021, September). A Recent Survey on 6G Vehicular Technology, Applications and Challenges. In: *2021 9th International Conference on Reliability, Infocom Technologies and Optimization (Trends and Future Directions) (ICRITO)* (pp. 1-5). IEEE.

[23] Onyema, E. M., Eucheria, N. C., Obafemi, F. A., Sen, S., Atonye, F. G., Sharma, A., & Alsayed, A. O. (2020). Impact of Coronavirus pandemic on education. *Journal of Education and Practice*, 11(13), 108-121.

[24] Marc, J. R. (2007). Book review: e-learning strategies for delivering knowledge in the digital age. Internet and Higher Education, 5, 185-188. https://doi.org/10.1002/pfi.4140410512.

[25] Fedynich, L. V. (2013). Teaching beyond the classroom walls: The pros and cons of cyber learning. *Journal of Instructional Pedagogies*, 13.

[26] Rahayu, R. P., & Wirza, Y. (2020). Teachers' Perception of Online Learning during Pandemic COVID-19. *Journal Penelitian Pendidikan*, 20(3), 392-406.

[27] Aithal, P. S., & Aithal, S. (2016). Impact of on-line education on higher education system. *International Journal of Engineering Research and Modern Education (IJERME),* 1(1), 225-235.

[28] Dani, R., Kukreti, R., Negi, A., & Kholiya, D. (2020). Impact of COVID-19 on education and internships of hospitality students. *International Journal of Current Research and Review*, 12(21), 86-94.

[29] Sari, Y. R., Endriani, D., & Rasmita, R. (2020). Tips and Tricks of Teaching Children at Home during Online Learning: Tips dan Triks Mengajar Anak di Rumah selama Pembelajaran Online. *Jurnal Laporan Abdimas Rumah Ilmiah*, 1(1), 55-61.

[30] Sansanwal, K., Shrivastava, G., Anand, R., & Sharma, K. (2019). Big data analysis and compression for indoor air quality. *Handbook of IoT and Big Data*, pp. 1-21.

[31] Gupta, A., Anand, R., Pandey, D., Sindhwani, N., Wairya, S., Pandey, B. K., & Sharma, M. (2021). Prediction of Breast Cancer Using Extremely Randomized Clustering Forests (ERCF) Technique: Prediction of Breast Cancer. *International Journal of Distributed Systems and Technologies (IJDST)*, 12(4), 1-15.

[32] Jain, S., Kumar, M., Sindhwani, N., & Singh, P. (2021, September). SARS-Cov-2 detection using Deep Learning Techniques on the basis of Clinical Reports. In: *2021 9th International Conference on Reliability, Infocom Technologies and Optimization (Trends and Future Directions)* (ICRITO) (pp. 1-5). IEEE.

[33] Sindhwani, N., Anand, R., Meivel, S., Shukla, R., Yadav, M. P., & Yadav, V. (2021). Performance Analysis of Deep Neural Networks Using Computer Vision. *Machine Learning*, 15, 17.

[34] Anand, R., Singh, J., Pandey, D., Pandey, B. K., Nassa, V. K., & Pramanik, S. (2022). Modern Technique for Interactive Communication in LEACH-Based Ad Hoc Wireless Sensor Network. In: *Software Defined Networking for Ad Hoc Networks* (pp. 55-73). Springer, Cham.

[35] Juneja, S., Juneja, A., & Anand, R. (2019, April). Reliability modeling for embedded system environment compared to available software reliability growth models. In: *2019 International Conference on Automation, Computational and Technology Management* (ICACTM) (pp. 379-382). IEEE.

Chapter 5

Evaluation of the SARS-COV-2 Omicron Variant's Features, Prospective Consequences, and Upcoming Prospects

Peter Ngulube[1], Binay Kumar Pandey[2,*], Digvijay Pandey[3], Vinay Kumar Nassa[4], A. Shaji George[5], Mukesh Kumar Gupta[6] and Pankaj Dadheech[6]

[1]Department of Biological Sciences, Immunology, Malawi University of Science and Technology, Malawi
[2]Department of Information Technology, College of Technology, Govind Ballabh Pant University of Agriculture and Technology, Pantnagar, India
[3]Department of Technical Education, IET, Dr A.P.J. Abdul Kalam Technical University, Lucknow, India
[4](AIDS) Department at Sharad Institute of Technology College of Engineering, Ichalkaranji (Kolhapur- Maharashtra-416121), India
[5]Department of Information and Communication Technology, Crown University, Int'l. Chartered Inc.(CUICI) Argentina Campus, South America
[6]Department of Computer Science & Engineering (NBA Accredited), Swami Keshvanand Institute of Technology, Management & Gramothan (SKIT), Jaipur, Rajasthan, India

Abstract

According to reports, the 2019 Corona-virus infection COVID-19 caused significant damage to world-wide demographic wellbeing. In South Africa, a new level severe respiratory syndrome coronavirus 2 (SARS-Cov-2) variant, B.1.1.529, was found awhile back, resulting in a

* Corresponding Author's Email: binaydece@gmail.com.

In: An Interdisciplinary Approach in the Post-COVID-19 Pandemic Era
Editors: D. Pandey, B. Kumar Pandey, N. Sindhwani et al.
ISBN: 979-8-88697-052-4

substantial increase in COVID-19 patients. Then On November 24, 2021, the World Health Organization outlined B.1.1.529, also known as omicron, as a variant under inspection. The Omicron variability must have been proclaimed as a variation of worry. This variant includes a huge number of modifications, particularly 15 inside the spike's receptor-binding region (RBD). The Omicron variant seems to be likewise similar to the preceding VOC Alpha, Beta, and Gamma variants, generating issues regarding virus infectiousness, pathogenicity, and immunity resistance. In this paper, the identification and features of the Omicron variant were discussed, contrasting the spiking alterations inside the five VOCs and discussing prospective avoidance and mitigation measures.

5.1. Introduction

This coronavirus infection (COVID-19) outbreak has now been spreading for more than two months. According to WHO figures, around 280 million proven incidents were reported, including over 6 million deaths [1]. The preliminary respiratory infection syndrome coronavirus 2 (SARS-Cov-2) outbreak occurred nearly at the end of 2019 [2-5], resulting in numerous variants. To prioritise tracking and research, WHO split such variants into three categories: variants of concern (VOCs), variants of interest (VOIS), and variants under monitoring (VUMS). The preceding four VOCS (b.1.617.2) seem to have been alpha (B.1.1.7), beta (B.1.351), gamma (P.1), and delta (b.1.617.2) [6]. They all culminated in a new epidemic and thousands of fatalities in more than one nation and region, if not the whole planet. On November 26, 2021, a novel variant, omicron (B.1.1.529), was discovered as one of the fifth variants of concern [7].

5.1.1. Data and Methodology

This research investigated the critical function of a SARS-CoV-2, Omicron Variant with concerns, features, possible impacts, as well as preventative measures throughout Africa. Studies have been considered provided they satisfy the basic guidelines: research which examines both the origin and effect of the SARS-CoV-2 Omicron variation on existing vaccinations, primary or secondary research, and also studies generally published in English.

The studies were excluded if they fulfilled the following exception criteria: Studies were not applicable to study subjects. Studies did not discuss the emergence and impact on vaccine efficacy of thevariant. For instance, primary research has been conducted on the emergence of an influence of an Omicron variation; investigations included pilot projects, editorials, letters, publications, study protocols, comments, and practical guides.

(a) Sources of Data and Search Strategy

On January 30, 2022, one author completed a comprehensive computer search in four databases; Google Scholar, PubMed, and Science Direct. SARS-CoV-2, Omicron Variant, Spike protein, Vaccine, and a search term 'SARS-CoV-2 Omicron variant' were the keywords used.To obtain the most narrowly relevant literature, Boolean operators were used in combination with refined search terms such as AND, OR, and NOT [8] on Science Direct and Google Scholar. A systematic search by the use of Medical Subject Headings (Mesh) was conducted on PubMed. Mesh searching was used to retrieve only literature relevant to the objectives of this systematic review. Other related articles were tracked down using the snowball method. All retrieved literature, including keywords, titles, abstracts, author's names, year of publication, and publishing journal, were exported to Mendeley software and Microsoft Excel sheets for analysis and reference management. In the first place, the whole literature was scanned for duplicates; hence, all duplicates were removed using the Mendeley reference manager. Two independent reviewers screened the titles and abstracts of the remaining studies. If the decisions of the two reviewers were unaltered, the study was included without deliberation in the second level. In the second step, the two reviewers separately examined the complete texts of the selected articles. In the discussion session, the reviewers reached a consensus on the outcome of a study in the case of disagreements.

(b) Data Extraction

Relevant data were collected based on the definition of the study's topic, inclusion and exclusion criteria, description of the methods used, outcomes, and statistical analysis.

5.2. Current COVID-19 Update in Africa

COVID-19 cases had been reported in 40 of Africa's 47 nations as of March 31, 2020, with a total of 3786 confirmed cases and 77 fatalities [9]. Almost all

African governments responded to the WHO's appeals and guidance by restricting and passing most shared boundaries, air and sea entries into and around their respective nations, while permitting essential critical activities to be carried out. Furthermore, many countries, including Tanzania, Algeria, Djibouti, Kenya, Morocco, Zambia, and many others, have put limits on social assemblies and domestic movement, whereas others, such as Rwanda and South Africa, have implemented additional lockdown restrictions [10].

South Africa and Rwanda are amongst the African nations that connected India, China, France, Italy, New Zealand, Spain, the United States, and Poland as the leading nations with the highest declared number of cases in the region just after the entire lockdown.It appears that the WHO and governments worked closely together, since countries made significant efforts to prevent this terrible epidemic by implementing WHO guidelines [3]. South Africa is the first known country with COVID-19 Omicron, but there are believes that it is also in the continents but not yet discovered [7]. The occurrence of the Omicron type was recorded in 22 African nations as of December 23, 2021. According to the Africa CDC, six African nations have recorded Omicron COVID-19 variant cases: Burkina Faso, Togo, Egypt, Kenya, Morocco, and Mauritius [11]. The African Union (AU) has cautioned that travel and entrance prohibitions imposed in response to the development of the Omicron COVID-19 variant restrict free movement of people and products, bearing immediate and serious consequences on African countries.

As of December 23, 2021, the confirmed COVID-19 case scenarios in Africa would have attained 9,259,813 with such a number of deaths as 226,536 [12]. According to WHO, the possibility of further global distribution of Omicron would be significant given the genetic variations that may confer immune flee capability as well as possibly transmissibility added benefit [12].

5.2.1. The Emergence of Omicronvariant in Africa

The very first confirmed omicron incident, as per WHO data, has been discovered inside a sample procured on November 9, 2021 [7]. On November 11, 2021, the first Omicron primers were obtained from Botswana samples [13]. Ever since exploration, the variant has seemed to be quickly disseminating.

As per the latest genome-wide study of 77 viral specimens taken throughout South Africa's Gauteng province from November 12 to 20 [14], all the analysed variants seem to have been B.1.1.529.In addition, the discovery of Omicron happens to coincide with such a growth in the total count of acknowledged COVID-19 cases throughout South Africa. Whenever the Omicron variant has been affirmed, the average number of COVID-19 case scenarios per day has risen from 280 to 800 [15].

Besides that, investigation into the beginnings of COVID-19 cases demonstrated that B.1.1.529 would have propagated all across Western Europe prior to the very first cases being identified throughout Southern Africa [16]. On November 24, 2021, the B.1.1.529 version was first mentioned by WHO. The WHO classified it as VUM and named it the Omicron diverse selection. In only two days, WHO categorised the Omicron variation as VOC, shattering the old history again for the shortest possible time between such a variant's restructuring from VUM to VOC and the mass popular worry.

Omicron was identified just a few days ago in Africa, and now it has migrated to the rest of the world. The occurrence of the Omicron was recorded in 22 African nations as of December 23, 2021 [11]. The African Union (AU) has cautioned that travel and entrance prohibitions imposed in response to the development of the Omicron COVID-19 variant restrict the free movement of people and products, bearing immediate and serious consequences for African countries [17].

This has seemed to indicate that perhaps the mutation has been continuing to spread to new regions and countries. According to phylogenetic analysis [18], the omicron variation progressed slightly earlier from those other SARS-Cov-2 variations than from one of the prior VOCS. The Omicron variant could be envisioned in immuno-suppressed people, but that could have progressed inside a non-human animal and has only recently been brought back into contact with humans [19].

As of December 23, 2021, there have been 9,259,813 COVID-19 cases throughout Africa, with 226,536 deaths [17]. Given the changes that may offer immunological escape potential and maybe transmissibility advantage, the risk of omicron spreading farther globally is significant, according to WHO.

5.3. Characteristics and Potential Ompacts of Omicronvariant

5.3.1. Clinical Features and Complications

According to studies, the median incubation period for omicron might be as little as 2–3 days, compared to 5 days for the original virus [20]. The much more prevalent side consequences of omicron do seem to be sniffles, sinus infections, migraines, tiredness (reasonable to serious), nasal congestion, as well as excessive sweating during the night [21]. Sickness, difficulty breathing, and odour loss are much less common in this version than in previous versions. Clinical symptoms may persist for several weeks before disappearing [22]. According to a study from South Africa, England, Scotland, Denmark, and Canada [23], people with omicron infections appeared to have a lower risk of hospitalization, catastrophic disease, and death rates than people with delta infectious diseases.

As a result, preliminary research suggests that omicron seems to be a relatively milder strain of SARS-CoV-2 than prior strains. The fact that, despite an increase in infectious diseases, hospital admissions, intensive care admissions, and death rates in both South Africa and the United Kingdom remained low [24], lends credence to the claim. Netcare, a private care institution in South Africa, recently posted an assessment that looked at the features of people who were admitted to a hospital during the four waves. Such waves were formed by the ancestor's viruses, beta, delta, and omicron, in that order [25].

The proportion of sick people needing oxygen therapy (17.6% in wave 4 vs. 74% in wave 3, p 0.001), and also the proportion of sufferers who required ventilatory support, both declined substantially. In ripple 4, 18.5 percent of people have been admitted to intensive care, especially in comparison to 29.9 percent throughout ripple 3 (p 0.001) [26]. Ripple 4 has seen a reduction in the average duration of stay from 7 to 8 days in earlier waves to three days. The mortality rate has gone up and down, with both 19.7% and 29.1% in waves 1 and 3, prior to actually dropping to 2.7 percent in waves 4 [27]. In 2019, coronavirus disease waves are depicted in Figure 5.1. Figure 5.2 depicts the proportion of sequential circumstances that protect against an outbreak.

days. 5.2.

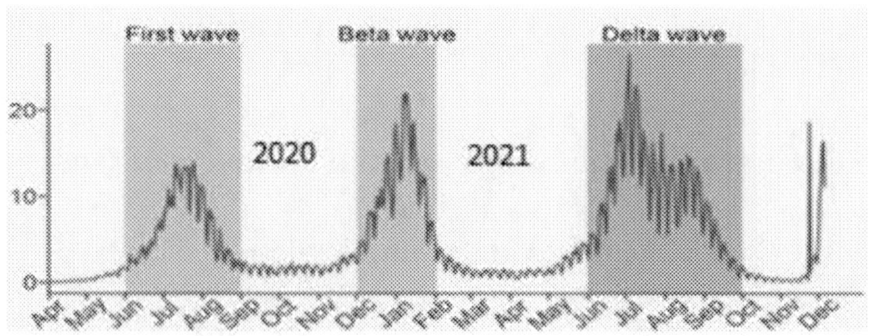

Figure 5.1. Coronavirus disease 2019 (covid-19) epidemic waves in South Africa.

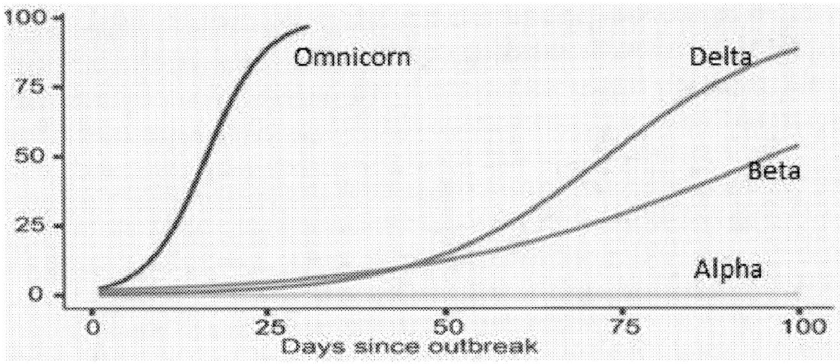

Figure 5.2. Omicron is growing faster in South Africa than other kinds of corcern (VOCs).

As of December 23, 2021, there have been 9,259,813 COVID-19 cases throughout Africa, with 226,536 deaths [17]. Given the changes that may offer immunological escape potential and maybe transmissibility advantages, the risk of omicron spreading farther globally is significant, according to WHO.

However, these were all preliminary findings, and the scene could very well start changing as days pass on. There has never been any data on how omicron affects sick people who are critically ill.

Moreover, even if the severity is minor, the checking ability as well as the medical system would've been overloaded [29-32].

It's also worth considering how omicron might affect the severity of long-term COVID. Finally, there is the cost of a significant increase in incidence on the international financial system, as well as increased awareness of other illnesses, social and cultural psychological wellness, and medical workers [19].

Virology

Omicron has genetic changes in common with alpha, beta, gamma, and delta variants, and also unknown genetic changes [19]. The characteristics of some of these mutations are currently being identified. It split into two lineages within a few months after it was created. The omicron's behaviour throughout cultured cells has always been enchanting [33]. According the investigators, Omicron, like some other seasonal coronaviruses but not SARS-CoV-2, has been hard to derive in VERO cells (genetically produced African green monkey kidney cells).

This has a lower likelihood of causing obvious damage (plaques) within those molecules [7].

In addition, omicron has a decreased ability for syncytia formation, which is prevalent in COVID-19 patients' lungs. Bronchi expand 10 times faster than delta (accounting for rapid dissemination), but lungs develop ten times slower [34], according to Hong Kong researchers. While the research is still in its early stages, these findings might explain why omicron is associated with a milder form of sickness [35].

Trans-missibility

Omicron is a highly infectious variant, with massively increasing times of 3.38 days (95 percent CI 3.18–3.61 days) and 2–2.5 days in South African and United Kingdom studies [17, 18], and then a very basic replication amount (R0) greater than 3.One such feature, which describes the fast expansion as well as dispersion of the notable delta variability.

5.4. Proposed Strategies for Prevention of Omicron Variant in Africa

5.4.1. Expanding the Availability of COVID-19 Vaccination

Although some believe that Omicron's rapid spread in South Africa heralds a new wave of pandemics around the world, the relevance of such a modified version as well as what it implies again for recent outbreak remain a mystery. In reality, the state of Omicron dissemination in South Africa may be significantly different from that in other countries. Just approximately 24 percent of the people in South Africa, for example, has received all of their vaccinations [36].

This figure is significantly lower than the global average vaccination rate of 42% [36]. This could hasten the propagation of omicron throughout South Africa, highlighting the critical need to increase the nation's immunization rates. Despite the fact that licenced covid-19 vaccines were less effective to the genetically changed viruses [21, 37], the results are very promising. Vaccines have been shown to prevent major illnesses, hospitalisation, and death [35].

Low antibody levels in SARS-Cov-2 infection occurs or vaccinated patients could aid in the development and selection of new SARS-Cov-2 variants.

Numerous studies have also found that serum neutralisation antibodies decrease significantly six months after immunization and also that supplemental vaccination with a precaution dose can indeed reinstate and possibly even improve vaccine efficacy [26]. Researchers believe that including an additional precautiondose of the COVID-19 vaccine in the immunizations will not only help to control Omicron infection but will also help to spread it.

5.4.2. Developing Vaccinations that are Variant-Specific

The occurrence of the Omicron variant was already linked to an increased risk of SARS-Cov-2 re-infection, particularly in South Africa, implying also that the Omicron variant may now be associated with a statistically significant power to inhibit protection from previous infection [26].

There is also a lot of curiosity about whether the current COVID-19 vaccinations could indeed safeguard against an Omicron version. According to the most recent studies, the COVID-19 vaccination provides less protection to the omicron form than other VOCS. Meanwhile, compared to the wild-type sars-cov2, vaccinated individuals' sera exhibited roughly 40 percent reduced neutralising powers against the omicron version.These findings showed that current covid-19 vaccinations may not be as efficient as other sars-cov-2 versions against the omicron form. In the future, more records on the usefulness of the presently offered COVID-19 vaccines would be required. Whereas the consequence of omicron spike genetic variations on vaccine efficacy would be unidentified, vaccines based on wild sars-cov-2 have been shown to be less effective across all prohibited variations of infectious diseases [30]. According to previous research [38], the vaccination based on mutant spike would already have resulted in a higher level of neutralising antibodies to the mutated viruses but only a low level of specific antibodies against wild-type sars-cov-2. Such findings highlight the importance of developing variant-specific vaccines based on mutant spikes, especially for omicron variability. As a consequence, scientists are exploring immunisations against the sars-cov-2 omicron modified version, focusing on the omicron version's altered spike. To protect omicron infectious disease and propagation, vaccine candidates depending on some other variants and containing each or even more omicron genetic changes could have been used. As per informal findings, Moderna must have created two multivalent vaccine candidates: mRNA-1273.211, which would be assumed to comprise numerous mutations discovered both in the omicron and beta variants, and mRNA-1273.212, which would be assumed to comprise a particular number of mutations encountered inside the omicron, beta, and delta variants [20] [39]. More experiments into the usefulness of such likely vaccines in contradiction of the omicron form has been required.

Conclusion and Future Recommendations

Omicron and others have dominated the world. This propagates faster and is more effective than older variations at trying to evade innate and vaccine-induced protection. Despite its low lethality, a public health effect could be substantial given the multitude of instances as well as spillover impacts on globalised trade. Booster immunizations are being used to limit the propagation of such a category in several advantaged nations, but their efficiency seems to be debatable. Time will indeed describe, regardless of whether Omicron would then start writing another deadly episode in the pandemic's historical record and whether it might very well end an outbreak by substituting the much more virulent variations.

References

[1] Coronavirus disease (COVID-19) [Internet]. [cited 2021 Dec 25]. Available from: https://www.who.int/emergencies/diseases/novel-coronavirus-2019.

[2] Peter SDN, Pandey D. Current Advancements on the Efficacy of Convalescent Plasma as a Potential Promising Treatment for COVID-19. *Asian J Adv Med Sci,* [Internet]. 2020; 2(3):7-12. Available from: https://mbimph.com/index.php/AJOAIMS/article/view/1719%0A.

[3] Pandey D, Gul R, Canete JJO, Rocha IANCN, Gowwrii G, Pandey BK, Rohana and Ngulube Peter SD. Mental Stress in Online Learning During the Pandemic : An Assessment of Learners ' Perception. 2021; 6(3):37-49.

[4] Jain S., Kumar M., Sindhwani N and Singh P. (2021, September). SARS-Cov-2 detection using Deep Learning Techniques on the basis of Clinical Reports. In *2021 9th International Conference on Reliability, Infocom Technologies and Optimization (Trends and Future Directions)* (ICRITO) (pp. 1-5). IEEE.

[5] Anand R., Sindhwani N and Saini A. (2021). Emerging Technologies for COVID-19. *Enabling Healthcare 4.0 for Pandemics: A Roadmap Using AI, Machine Learning, IoT and Cognitive Technologies*, 163-188.

[6] Tracking SARS-CoV-2 variants [Internet]. [cited 2021 Dec 25]. Available from: https://www.who.int/en/activities/tracking-SARS-CoV-2-variants.

[7] Karim SSA, Karim QA. Omicron SARS-CoV-2 variant: A new chapter in the COVID-19 pandemic. *Lancet,* 2021 Dec; 398(10317):2126-8.

[8] Harari MB, Parola HR, Hartwell CJ, Riegelman A. Literature searches in systematic reviews and meta-analyses: A review, evaluation, and recommendations. Vol. 118, *Journal of Vocational Behavior*. Academic Press Inc.; 2020. p. 103377.

[9] Situation Report-71 HIGHLIGHTS. [cited 2021 Dec 30]; Available from: https://www.who.int/dg/speeches/detail/who-director-general-.

[10] Liu J, Liao X, Qian S, Yuan J, Wang F, Liu Y,Wang Z, Wang F-S, Liu L and Zhang Z. Community transmission of severe acute respiratory syndrome Coronavirus 2, Shenzhen, China, 2020. *Emerg Infect Dis,* 2020 Jun 1; 26(6):1320-3.

[11] Omicron spreads but severe cases remain low in South Africa|WHO|Regional Office for Africa [Internet]. [cited 2021 Dec 25]. Available from: https:// www.afro.who.int/news/omicron-spreads-severe-cases-remain-low-south-africa.

[12] Coronavirus (COVID-19)|WHO|Regional Office for Africa [Internet]. [cited 2021 Dec 30]. Available from: https://www.afro.who.int/health-topics/coronavirus-covid-19.

[13] Enhancing Readiness for Omicron (B.1.1.529): Technical Brief and Priority Actions for Member States.

[14] Graham F. Daily briefing: Omicron coronavirus variant puts scientists on alert. *Nature,* 2021 Nov 26.

[15] Graham F. Daily briefing: Omicron coronavirus variant puts scientists on alert. *Nat,* 2021 [Internet]. 2021 Nov 26 [cited 2021 Dec 25]; Available from: https:// www.nature.com/articles/d41586-021-03564-6.

[16] Omicron COVID variant was in Europe before South African scientists detected and flagged it to the world - CBS News [Internet]. [cited 2021 Dec 25]. Available from: https://www.cbsnews.com/news/omicron-variant-covid-in-europe-netherlands-before-alert-raised/.

[17] COVID-19 News – Africa CDC [Internet]. [cited 2021 Dec 25]. Available from: https://africacdc.org/covid-19-news/.

[18] Kupferschmidt K. Where did 'weird' Omicron come from? *Science,* (80-). 2021 Dec 3; 374(6572):1179-1179.

[19] Kumar S, Thambiraja TS, Karuppanan K, Subramaniam G. Omicron and Delta Variant of SARS-CoV-2: A Comparative Computational Study of Spike Protein. *J Med Virol*, [Internet]. 2021 Dec 15 [cited 2021 Dec 25]; Available from: https:// onlinelibrary.wiley.com/doi/10.1002/jmv.27526.

[20] Choi A, Koch M, Wu K, Chu L, Ma LZ, Hill A, Nunna N, Huang W, Oestreicher J, Colpitts T, Bennett H, Legault H, Paila Y, Nestorova B, Ding B, Montefiori D, Pajon R, Miller JM, Leav B, Carfi A, McPhee R and Edwards DK. Safety and immunogenicity of SARS-CoV-2 variant mRNA vaccine boosters in healthy adults: An interim analysis. *Nat Med*, 2021 Nov 1; 27(11):2025-31.

[21] Martínez-Baz I, Miqueleiz A, Casado I, Navascués A, Trobajo-Sanmartín C, Burgui C, Guevara M, Ezpeleta C and Castilla J. Effectiveness of COVID-19 vaccines in preventing SARS-CoV-2 infection and hospitalisation, Navarre, Spain, January to April 2021. *Eurosurveillance*, 2021 May 27; 26(21).

[22] Volz E, Hill V, McCrone JT, Price A, Jorgensen D, O'Toole Á, Southgate J, Johnson R, Jackson B, Nascimento FF, Rey SM, Nicholls SM, Colquhoun RM, da Silva Filipe A, Shepherd J, Pascall DJ, Shah R, Jesudason N, Li K, Jarrett R, Pacchiarini N, Bull M, Geidelberg L, Siveroni I, Consortium C-U, Goodfellow I, Loman NJ, Pybus OJ, Robertson DL, Thomson EC, Rambaut and Connor TR. Evaluating the Effects of SARS-CoV-2 Spike Mutation D614G on Transmissibility and Pathogenicity. *Cell,* 2021 Jan 7; 184(1):64-75.e11.

[23] Nunes B, Rodrigues AP, Kislaya I, Cruz C, Peralta-Santos A, Lima J, Leite PP, Sequeira D, Dias CM and Machado A. mRNA vaccine effectiveness against COVID-19-related hospitalisations and deaths in older adults: A cohort study based on data linkage of national health registries in Portugal, February to August 2021. *Eurosurveillance,* 2021 Sep 23; 26(38).

[24] Yan R, Zhang Y, Li Y, Xia L, Guo Y, Zhou Q. Structural basis for the recognition of SARS-CoV-2 by full-length human ACE2. *Science,* (80-). 2020 Mar 27; 367(6485):1444-8.

[25] Korber B, Fischer WM, Gnanakaran S, Yoon H, Theiler J, Abfalterer W, Hengartner N, Giorgi EE, Bhattacharya T, Foley B, Hastie KM, Parker MD, Partridge DG, Evans CM, Freeman TM, de Silva TI, Sheffield COVID-19 Genomics Group, McDanal C, Perez LG, Tang H, Moon-Walker A, Whelan SP, LaBranche CC, Saphire EO and Montefiori DC. Tracking Changes in SARS-CoV-2 Spike: Evidence that D614G Increases Infectivity of the COVID-19 Virus. *Cell,* 2020 Aug 20; 182(4):812-827.e19.

[26] Pulliam JRC, Schalkwyk C van, Govender N, Gottberg A von, Cohen C, Groome MJ, Dushoff J, Mlisana K and Moultrie H. Increased risk of SARS-CoV-2 reinfection associated with emergence of the Omicron variant in South Africa. *medRxiv,* [Internet]. 2021 Dec 2 [cited 2021 Dec 25]; 2021.11.11.21266068. Available from: https://www.medrxiv.org/content/10.1101/2021.11.11.21266068v2.

[27] Yuan M, Liu H, Wu NC, Wilson IA. Recognition of the SARS-CoV-2 receptor binding domain by neutralizing antibodies. *Biochem Biophys Res Commun,* 2021 Jan 29; 538:192-203.

[28] He X, Hong W, Pan X, Lu G, Wei X. SARS-CoV-2 Omicron variant: Characteristics and prevention. *MedComm*, [Internet]. 2021 Dec 1 [cited 2021 Dec 25]; 2(4):838-45. Available from: https://onlinelibrary.wiley.com/doi/full/10.1002/mco2.110.

[29] Leung K, Shum MHH, Leung GM, Lam TTY, Wu JT. Early transmissibility assessment of the N501Y mutant strains of SARS-CoV-2 in the United Kingdom, October to November 2020. *Eurosurveillance*, 2020 Dec 1; 26(1).

[30] Jain S, Sindhwani N, Anand R and Kannan R. (2022). COVID Detection Using Chest X-Ray and Transfer Learning. In *International Conference on Intelligent Systems Design and Applications*, (pp. 933-943). Springer, Cham.

[31] Meivel S, Sindhwani N, Anand R, Pandey D, Alnuaim AA, Altheneyan AS, Jabarulla MY and Lelisho ME. (2022). Mask Detection and Social Distance Identification Using Internet of Things and Faster R-CNN Algorithm. *Computational Intelligence and Neuroscience*, 2022.

[32] Babu SZD, Pandey D and Sheik I. (2020). Acts Of COVID19. *International Journal of Advanced Engineering*, 3, 2457-0397.

[33] Pilishvili T, Gierke R, Fleming-Dutra KE, Farrar JL, Mohr NM, Talan DA, Krishnadasan A, Harland KK, Smithline HA, Hou PC, Lee LC, Lim SC, Moran GJ, Krebs E, Steele MT, Beiser DG, Faine B, Haran JP, Nandi U, Schrading WA, Chinnock B, Henning DJ, Lovecchio F, Lee J, Barter D, Brackney M, Fridkin SK, Marceaux-Galli K, Lim S, Phipps EC, Dumyati G, Pierce R, Markus TM, Anderson DJ, Debes AK, Lin MY, Mayer J, Kwon JH, Safdar N, Fischer M, Singleton R, Chea N, Magill SS, Verani JR and Schrag SJ. Effectiveness of mRNA COVID-19 Vaccine among U.S. Health Care Personnel. N Engl J Med. 2021 Dec 16;385(25):e90.

[34] Dorothee Meyer E, Sandfort M, Bender J, Matysiak-Klose D, Dörre A, Bojara G, Beyrer K and Hellenbrand W. Two doses of the mRNA BNT162b2 vaccine reduce severe outcomes, viral load and secondary attack rate: evidence from a SARS-CoV-2 Alpha outbreak in a nursing home in Germany, January-March 2021. *medRxiv*, [Internet]. 2021 Sep 23 [cited 2021 Dec 25]; 2021.09.13.21262519. Available from: https://www.medrxiv.org/content/10.1101/2021.09.13.21262519v1.

[35] Andrews N, Tessier E, Stowe J, Gower C, Kirsebom F, Simmons R, Gallagher E, Chand M, Brown K, Ladhani SN, Ramsay M and Bernal JL, Vaccine effectiveness and duration of protection of Comirnaty, Vaxzevria and Spikevax against mild and severe COVID-19 in the UK. *medRxiv,* [Internet]. 2021 Oct 6 [cited 2021 Dec 25]; 2021.09.15.21263583. Available from: https://www.medrxiv.org/content/10.1101/2021.09.15.21263583v2.

[36] There are far more mutations than Delta, which has been detected in Hong Kong; is the Omicron mutant strain a "super strain"? [Internet]. [cited 2021 Dec 25]. Available from: https://mp.weixin.qq.com/s/ccms2rTGGxwSGs2VzSCHvg.

[37] Martínez-Baz I, Trobajo-Sanmartín C, Miqueleiz A, Guevara M, Fernández-Huerta M, Burgui C, Casado I, Portillo ME, Navascués A, Ezpeleta C and Castilla J. Product-specific COVID-19 vaccine effectiveness against secondary infection in close contacts, Navarre, Spain, April to August 2021. *Euro Surveill,* 2021 Sep 1; 26(39).

[38] He C, Yang J, He X, Hong W, Lei H, Chen Z, Shen G, Yang L, Li J, Wang Z, Song X, Wang W, Lu, G and Wei X. A bivalent recombinant vaccine targeting the S1 protein induces neutralizing antibodies against both SARS-CoV-2 variants and wild-type of the virus. *MedComm,* 2021 Sep; 2(3):430-41.

[39] Moderna Announces Significant Advances Across Industry-Leading mRNA Portfolio at 2021 R&D Day|BioSpace [Internet]. [cited 2021 Dec 25]. Available from: https://www.biospace.com/article/releases/moderna-announces-significant-advances-across-industry-leading-mrna-portfolio-at-2021-r-and-amp-d-day/.

Chapter 6

Impact on Various Sectors during COVID-19 Crisis

Ritu Singh[1,*], Rekha Narang[1,†], Prerna Dawar[1,‡], A. Shaji George[2,§] and Parbhash[1,‖]

[1]Geeta University, Panipat, Haryana, India
[2]Department of Information and Communication Technology, Crown University, Int'l. Chartered Inc.(CUICI) Argentina Campus, Santa Cruz Province, Argentina, South America

Abstract

COVID-19 has changed the lives of everyone for the worse in the world we live in. Even after two years, every part of the world is still being affected by economic downturns because of unemployment, downfalls in many businesses, and the closure of people in their homes. It has not only affected human health but also has an impact on the global economy that leads to job losses, business disruptions and makes us look back on one of the worst times ever for people on earth. Nearly all industries are declining sharply in their businesses, and the impact of this pandemic is so great that it predicts worse times to come. The present chapter has focused on the various industries which are most affected by the COVID-19 that has led to job losses and the downfall of the global economy.

Keywords: COVID-19, Indian economy, industries fallout, recession

* Corresponding Author's Email: ritulohra1412@gmail.com.
† Corresponding Author's Email: dr.narang12@gmail.com.
‡ Corresponding Author's Email: prernadawarsaluja@gmail.com.
§ Corresponding Author's Email: drashajigeorge@gmail.com.
‖ Corresponding Author's Email: sahprabhash7890@gmail.com.

In: An Interdisciplinary Approach in the Post-COVID-19 Pandemic Era
Editors: D. Pandey, B. Kumar Pandey, N. Sindhwani et al.
ISBN: 979-8-88697-052-4

1. Introduction

The coronavirus disease began in China in late 2019. In any case, by mid-2021, the whole world was feeling the effects of this pandemic [1-3]. The start of the COVID-19 epidemic is a major blow to the Indian economy. It was in a precarious position at that time before the COVID-19 strike. When the Indian economy is assumed to be one person, the payoff for 2020-2021 and 2021–2022 is usually not the same as it was in 2019-2020, according to the latest World Bank cages. There is a big, amazing, and financially abusive event to come. These two systems and management issues must embrace a critical role to reduce this so that all equals and negatives can be destroyed, even refunds. Thanks to huge profits and clever chains, the economy is likely to rise to prominence in the long run. The magnitude of the financial impact depends on the degree of success during the closure as well as the extent to which the situation has spread once the closure has been removed.

Most organisations that plan a certain way to persevere in the COVID-time financial framework will want to grow successfully. When the world saw this epidemic, there was anticipation of a buy-out of tourism and the tourism industry. Moving objects, advanced objects, metals, and object positions in the same way noted the height. Organizations reconsidered their auction strategy without efficiency or profit enhancement.

Collaborative companies guarantee a weighty balance simply because the development of the system will no longer find consumers around profitability and development will be a new normal environment. The closure of the country happened while the economy was struggling. Trade in all regions was tested for its impact. In addition to the import and charge businesses, another large pay generator that received an impact was the movement business. Indian shops were full of chaos. Also, various sectors, such as agriculture, the primary sector, and the tertiary education sector, have not yet been released on their impact. There were no jobs available for agricultural purposes in different provinces. The closure of Lockdown has made it difficult for farmers to make their products available for purchase in the commercial sector. India's free-cut piece, which is the backbone of its economy, will be even more difficult when it comes to money-related jobs reaching the bottom line. These barriers to business activities and open parties tend to contribute to the development of the family without moving. The uncontrolled spread of COVID-19 in the suburbs and in the regions has had a devastating effect on almost the entire world and created a significant financial threat [4]. As COVID-19 is rapidly spreading in India, policymakers are concerned about how to tackle the disease, which is seriously damaging the economy [5]. Policymakers should plan for the long-term difficulties and openness that can arise when a problem is wrapped up.

The COVID epidemic has already devastated India's economy. This feature was another problem with the closure of the sales area in order to address the existing difficulties. However, organisations providing hardware components like Paytm and Google Pay have shown some degree of benefit in this regard. The Indian economy had declined for several months when the Indian government announced the lockdown in March 2020. The COVID-19 and the closure of several facilities have also had a significant impact on the economy of the country. The impact of the coronavirus in India financially has, in general, been considered dangerous. As indicated by the Department of Statistics, Indian development for the last quarter of the 2020 financial year fell by 3.1%.

Lockdown has placed an unprecedented burden on the flexible chains of key products, and as a result, a large portion of Indian organizations focuses on the development of just the right things, as well as the suspension of any remaining activities, thereby preventing down the development diagram.

2. Influence on the Travel Industry

During COVID-19, more than 100 countries imposed exit restrictions, which impacted on the mobility business [6, 7]. The Indian tourism industry had a huge loss of Rs. 1.25 trillion in the 2020 plan as a result of the closure of accommodation and flight suspensions. During the months of April and June, the tourism industry in India relied on booking a salary loss of Rs. 69,400 crores, which means a yearly shortage of 30%. In addition, India's tourism industry has become an important partner in the country's GDP. This unit even created a large business base. So, then, in mid-March, significant landmarks in India began to close, and news of the airstrip began to make changes, all in all, beginning to predict the worst fate of the country's moving business.

3. Impact on the Gold Industry

The COVID-19 pandemic has devastated the economy of India. In a similar vein to various businesses, the jewellery business has also been hit hard by impressions as well as discounted or allowed events, exhibitions, presentations, and weddings for several months [8]. People also haven't spent money on any luxury items but have become aware and invested money on hygienic or basic food items. The diamond and jewellery industries have lost their brilliance. As a precautionary measure, the organisation has encouraged the residents to lower their marriage limits and try to find essential products during the closure. The business also lost its splendour to festivals, which collapsed during the lockdown and again, lost an unprecedented opportunity to see high-level preparations. While organisations are still facing difficulties, the effect of COVID-19 on the gold business is obviously very strong. Therefore, financial regulators in this period are experiencing major disasters in the Indian market.

4. Impact on Technology Sector

During the global closure, top-renowned software companies such as TCS, Infosys, and HCL were affected as equipment consumption decreased in the United States and Europe. Securities of HDFC expect about a 2-7% decline in the IT industry and a reduction in revenue as organisations assess the global economic downturn because of the COVID-19 pandemic. According to ICICI Direct, there was pressure testing, reduced revenue, and customer termination due to low market spending. According to Global Data reports, IT services were hit hard by the COVID-19 era. In addition to this, all technologies, media, the telecom industry, and related areas had negative effects during the COVID-19 era [9].

5. Impact on Stock Markets

Shares and Securities Markets are deteriorating globally to record the decline and impact of global financial conditions. Dow Jones having the result of -22.29% per month and -0.08% per year building represents the cases increased in the US [10]. It was unreasonable to expect a rapid financial escalation back to the current outcome of COVID-19 pandemic. Despite this fact that the financial crisis was inevitable, there is a reflection on the hard work of financial experts and public banks, in order to minimize the impact and significant financial suspensions. The issue in the current context is how to quickly and effectively meet the overall challenge of prosperity. The 2020 year was set to fall sharply in all parts of the world and in all stages of life. However, it depends on how quickly the epidemic was eradicated, as well as the selection of government-managed managers who were taking advantage of their economy. That's when the epidemic broke out with the distributed features returning to the business once and for all economically, currency trading will begin to move at a slower pace, and recovery will be faster than expected. The great thing about the market is that whether it is repairs or upgrades, these two categories make the trade of value or defense attractive and worth taking the openings.

One study stated that whenever the market rises somewhere near 20%, it takes 536 normal days for recovery. In any case, due to the COVID-19 crisis, Interest rates at various national banks in different countries declined in borrowing costs to reduce the effect on banks, allowing for less cost effective and a more cost-effective comfort relief to help the economy as well. Global markets intensified to some extent after the United States Senate passed a $ 2 trillion Coronavirus aid bill for helping the workers to identify in more organized manner.

6. Impact on Service/Jobs

Unemployment cases in the US are growing rapidly as the number of Americans applying for jobs reaches a record high, marking the decade-end extension in the largest economy in the world. Preliminary results indicate that the business shortfall is considered to be the worst effect of a crisis with low financial development and rising volatility, which can have a long-term impact. The lockdown situation in all nations has been the most glorious act ever. In any case, the assessment reveals an impact on employment during the closure and should not be construed as a long-term misfortune for the employee. This can create financial instability among the employees [11]. A large number of employees have the option to return to work after a lockdown closure. However, the facts confirm that they had a critical number and had no option to reverse their conditions. For example, hard-working workers previously moved to their primary residences. In this case, the free workers were the weakest because of the dreamy nature of their job and the daily wage portion, which is highly promoted in the development sector. In line with these trends, this group of truly basic agents, unemployed or unemployed people, talented professionals, and young financiers who may be quiet at home or returning to their homes or living in secluded homes, will undoubtedly not be able to so that they can resume their duties when the closing period expires. Incorporated health measures such as social removal, subsequent contact, and the ability to thrive over half of the workplace and market in the same way can affect a business employee's relationship, thus, being a major departure from simple business as usual.

7. Impact on Manufacturing Industry

The development of goods, as well as power and people, has come to a sudden halt. Micro, small, and medium enterprises are just a variety of smart businesses that stop working, earn less interest, and withdraw applications regardless of how you look at them, without worrying about the core business [12]. However, their requests due to significant obligations from idle businesses will not be met. The collection business is affected by different perspectives because of the Corona fact. It takes a certain amount of time to demonstrate, like a few traders, seeing the opportunity to reap the benefits of the lack of performance with a complete set of donated components. A growing number of agents are stopping work due to government directives, reducing exercise size, which has the following impact on quality, cost, and volume. Significant delays in banking operations, fewer working hours, delayed book lines, and excessive problems result in agreed-upon trade, which is related to fundraising. Providers to major creators start under pressure to show what else is starting to come out, and play safe to find their tendencies, as their ability to tolerate weaknesses is much lower than their larger clients. Eventually, because of all these issues, the end customer begins again to accept the unwanted purchase and withdraws from the used premises, postponing his requests.

8. Impact on Oil Industry

The demand for oil had declined when most of the world was in a state of lockdown. Outside of COVID, there is a line going on between OPEC's oil production and Russia's, which also shows a drop in global oil costs.

The Indian government earns most of its salary from bonds issued by a large portion of 90% starting with oil purchases. It is interesting to note that the cost of retailers has not received a discount since public officials subsidised their costs. As India imports more than 80% of its oil consumption, lowering oil costs will likely lower CAD in terms of the economy [13]. The current assets related to assets in CAD can now be used to maintain funding to reduce the COVID-19 portion. Indian oil companies in the E&P sector like ONGC and Oil India could face the dire consequences of considering the nature of their products at a lower cost in advance. Retailers such as HPCL, Reliance, in addition to IOCL, may see adjusted edges in nearby areas where interest arrives late. On the downside, if Indian organisations are able to manage their stores, this is a good opportunity to buy and hold oil for some time as a result. At that point, when the lockdown closure is closed, government officials may face extended difficulties in reducing fuel costs for consumers. Currently, a large part of the population is still at home in this closed state. From now on, it shows the burning of billions to billions in their exhibitions to expand the sales of their management systems by thinking about a variety of common advertising strategies and understanding the net of customer actions in the United States.

9. Impact on Hotel Industry

The hotel industry is one of the most influential industries during the epidemic, as demand has dropped to very low. Due to International travel advice, Visa suspension and Section 144 imposition (ban on public gatherings), Industrial workers have been severely affected by the COVID-19 epidemic. Not only have more people in the industry lost their jobs, but they have also seen significant changes in the way jobs should be developed. The Impact of Employees in the Hospitality examines the different ways of employees' recent and background details. The story begins with a discussion of how the hospitality industry has suffered in more countries than others, such as India, Europe and China. The United States has also experienced unprecedented liquidation and a severe business collapse over the past six months and will be even worse if there is a pandemic. The industry has an estimated 8.3 million positions in the United States alone, including accommodation, incinerator burns, an extended shopping network. However, as accommodation prices fall sharply, this creates major employment problems. Since the introduction of Coronavirus in the United States, accommodation has lost more than $10 billion in room revenue until the first seven days of April 2020, and will rise to $ 3.5 billion if the situation worsens. The situation in the tourism industry is much worse than in 2001 and 2007. In 2009, the United States suffered. The situation could be even worse if you consider the recent demographics, the loss of up to 6.5 million positions in the United States Hotel industry. As a result, many positions are declining in a remote country. One can expect the impact of COVID-19 on the entire tourism industry [14].

10. Impact on Market Industry

Marketing is related to business communication with customers and various organizations, primarily B2C and B2B. Currently, most people are at home in a closed state. It therefore marks a complete overhaul of millions to millions in their advertising to build their management sales through a variety of common marketing strategies and understanding the needs of customers and financial management strategies.

11. Impact on Education Industry

In the lockdown, schools and colleges were closed, and students attended the classes through various means, such as online-based home rooms and radio projects. However, it was good that, on the other hand, there were a lot of students who did not have the resources to go to online-based classes and they tolerated a ton. Many students were trying to find the necessary resources for online classes [15].

Professional teachers on board, chalk, textbooks, and in the classroom are new to this computer instruction, yet they embraced new techniques and treated them like stars to help students in the current situation. In any case, on the negative side, most teachers were looking for a special job to help their families. Educated caregivers supported their children throughout.

There are Indian students coming to school as they can get food. A wonderful breakfast party has helped many children who could not carry their food from home to get their own food, but in lockdown, students did not have enough food to survive. There was a postponement or withdrawal of the examination, which created tension in some undergraduates, and there was no space for the education system.

The vast majority of school-going teens are associated with their children's work to help their families. The pandemic has affected students and the foundation of the low school system and schools, which has led to the closure of the same thing. There was both good and bad news happening around us in the middle of the COVID-19 [16]. Innovation prepares for training, thus helping students and faculty communicate all the purposes and objectives of web-based learning halls, online classes, computer exams, etc. However, the sad reality is that it is not accessible to most students across the country.

12. Impact on Pharmaceutical Industry

China manufactures 70% of Indian pharmaceutical active ingredients (APIs). Active pharmaceutical ingredients (APIs) are essential in any country in the pharmaceutical world. As COVID-19 cases in India increase to 101K, drugs are growing in demand because there are not enough APIs in place, so the market will prove to be at higher prices.

Conclusion

The impact of the coronavirus on a variety of activities is unbelievable, and it will continue to disrupt the global economy till its antibody is expected to be vigilant and given to at least 50% of the population [17]. The pandemic caused a shortage of millions of jobs in all businesses and had a much bigger impact than the decline in 2008. Businesses have been hit hard in accommodation, tourism, transportation, oil, entertainment, land, and development and marketing. In the US alone, the hotel industry is expected to lose $3.5 billion/week and about 6.5 million positions in a combined 8.3 million inn operations, especially in view of the development of COVID-19 from now on. Therefore, the future awaits the continuation of some of the worst downturns ever [18, 19]. Some sort of miracle or supernatural event could eventually put an end to the pandemic's effect.

References

[1] Wu, F, Zhao, S., Yu, B., Chen, YM. Wang, W., Song, Z. G., Hu, Y., Tao, Z. W., Tian, J. H., Pei, Y. Y., Yuan, M. L. A new coronavirus associated with human respiratory disease in China. *Nature,* 1, 1-5(2020).

[2] Anand, R., Sindhwani, N., & Saini, A. (2021). Emerging Technologies for COVID-19. *Enabling Healthcare 4.0 for Pandemics: A Roadmap Using AI, Machine Learning, IoT and Cognitive Technologies*, 163-188.

[3] Meivel, S., Sindhwani, N., Anand, R., Pandey, D., Alnuaim, A. A., Altheneyan, A. S., ... & Lelisho, M. E. (2022). Mask Detection and Social Distance Identification Using Internet of Things and Faster R-CNN Algorithm. *Computational Intelligence and Neuroscience*, 2022.

[4] Fernandes, N., *Economic Effects of Coronavirus Outbreak (COVID-19) on the World Economy*. (2020).

[5] McKibben, W. J., Fernando, R. The Global Macroeconomic Impacts of COVID-19: Seven Scenarios. *CAMA Working Paper* No. 19/2020. (2020).

[6] Khongsai, L., Anal, T. S., As, R., Shah, M., & Pandey, D. (2021). Combating the spread of COVID-19 through community participation. *Global Social Welfare*, 8(2), 127-132.

[7] Bhati, A. S., Mohammadi, Z., Agarwal, M., Kamble, Z., & Donough-Tan, G. (2021). Motivating or manipulating: The influence of health-protective behaviour and media engagement on post-COVID-19 travel. *Current Issues in Tourism*, 24(15), 2088-2092.

[8] Jowitt, S. M. (2020). COVID-19 and the global mining industry. *SEG Discovery*, (122), 33-41.

[9] Parthiban, K., Pandey, D., & Pandey, B. K. (2021). Impact of SARS-CoV-2 in online education, predicting and contrasting mental stress of young students: a machine learning approach. *Augmented Human Research*, 6(1), 1-7.

[10] Gormsen. N. J., Koijen. R. S. J. Coronavirus: Impact on Stock Prices and Growth Expectations. Chicago: University of Chicago. *Becker Friedman Institute for Economics Working Paper* No. 2020-22. (2020).

[11] Fetzer, T., Hensel, L., Hermle J. Roth, C. Coronavirus Perceptions and Economic Anxiety. *ArXiv Preprint arXiv:* 2003.03848. (2020).

[12] Jain, S., Kumar, M., Sindhwani, N., & Singh, P. (2021, September). SARS-Cov-2 detection using Deep Learning Techniques on the basis of Clinical Reports. In *2021 9th International Conference on Reliability, Infocom Technologies and Optimization (Trends and Future Directions) (ICRITO)* (pp. 1-5). IEEE.

[13] Hanieh, A. (2021). COVID-19 and global oil markets. *Canadian Journal of Development Studies/Revue canadienne d'études du développement*, 42(1-2), 101-108.

[14] Lai, I. K. W., & Wong, J. W. C. (2020). Comparing crisis management practices in the hotel industry between initial and pandemic stages of COVID-19. *International Journal of Contemporary Hospitality Management*.

[15] Pandey, D., Ogunmola, G. A., Enbeyle, W., Abdullahi, M., Pandey, B. K., & Pramanik, S. (2021). COVID-19: A framework for effective delivering of online classes during lockdown. *Human Arenas*, 1-15.

[16] Singh, H., Rehman, T. B., Gangadhar, C., Anand, R., Sindhwani, N., & Babu, M. (2021). Accuracy detection of coronary artery disease using machine learning algorithms. *Applied Nanoscience*, 1-7.

[17] Jain, S., Sindhwani, N., Anand, R., & Kannan, R. (2022). COVID Detection Using Chest X-Ray and Transfer Learning. In *International Conference on Intelligent Systems Design and Applications* (pp. 933-943). Springer, Cham.

[18] Gupta, A., Anand, R., Pandey, D., Sindhwani, N., Wairya, S., Pandey, B. K., & Sharma, M. (2021). Prediction of Breast Cancer Using Extremely Randomized Clustering Forests (ERCF) Technique: Prediction of Breast Cancer. *International Journal of Distributed Systems and Technologies (IJDST)*, 12(4), 1-15.

[19] Juneja, S., Juneja, A., & Anand, R. (2020). Healthcare 4.0-Digitizing Healthcare Using Big Data for Performance Improvisation. *Journal of Computational and Theoretical Nanoscience*, 17(9-10), 4408-4410.

Chapter 7

COVID-19 and Post-COVID Effect in India: A Critical Analysis

Rajani Srivastava[1,*], Dharmesh Srivastava[2,†] and Sadhana Gond[1,‡]

[1]Institute of Environment & Sustainable Development, RGSC, Banaras Hindu University, Uttar Pradesh, India

[2]S R Institute of Management and Technology, Lucknow, India

Abstract

India is in the second position in terms of population and the third-worst-affected country in the world in the total number of confirmed COVID-19 cases. COVID-19 was marked by low case fatality rates and strong recovery rates in India, as well as an increase in public-private partnerships in the health sector. Technological advancements have aided in the containment of this epidemic. The virus's growth rate was halted by an early lockdown policy, however, super events are increasing infection rates due to the breakdown of COVID protocol behavior. This COVID-19 pandemic is caused by the severe acute respiratory syndrome coronavirus 2 (SARS-CoV-2). It contributes to worldwide healthcare issues and overstretched healthcare resources. As the people recovering from this disease, it is of utmost importance to establish knowledge of healthcare issues surrounding them. COVID-19 is a multi-organ disease and there are reports of permanent and prolonged effects of this disease. So it is very important to understand COVID-19, the post-COVID effect and the status of the administration and individuals dealing with this

* Corresponding Author's Email: rajani.srivastava25@gmail.com.
† Corresponding Author's Email: dharmesh2809@gmail.com.
‡ Corresponding Author's Email: sadhanagond7@gmail.com.

In: An Interdisciplinary Approach in the Post-COVID-19 Pandemic Era
Editors: D. Pandey, B. Kumar Pandey, N. Sindhwani et al.
ISBN: 979-8-88697-052-4

situation. The present chapter discusses the knowledge regarding COVID-19, various post-COVID-19 symptoms reported all over the world, especially in India in terms of professional status, financial status, mental status and status of the government and people dealing with this pandemic situation.

Keywords: COVID-19, post-COVID effect, pandemic, India, management

1. Introduction

Severe acute respiratory syndrome coronavirus 2 (SARS-CoV-2) is the pathogen that causes acute coronavirus disease (COVID-19) [1-4]. COVID-19 virus belongs to the *Coronavirinae* subfamily, part of the *Coronaviridae* family, and possesses a single-stranded ribonucleic acid structure. It has a well-defined structure consisting of 14 binding residues that directly interact with human angiotensin-converting Enzyme 2 [5, 6]. In December 2019, the first time this novel kind of virus was identified in Wuhan, the People's Republic of China, where a bunch of cases of 'viral pneumonia' was reported [7]. After that, more than 175 million people have been reported infected globally within a year, and have caused sickness and death at an exceptional scale worldwide [6, 8]. This coronavirus travelled all three stages, firstly in the form of 'Outbreak,' after that 'Epidemic' and then 'Pandemic.' WHO defines a Pandemic as the "worldwide spread of a new disease." COVID-19 was declared as PHEIC (Public Health Emergency of International Concern) on 30th January 2020 which means it is mandatory to report to WHO [9]. India is in the second position in terms of population and the third-worst-affected country in the world in total number of confirmed COVID-19 cases. COVID-19 was marked by low case fatality rates and strong recovery rates in India and witnessed a spread of public-personal partnership inside the health zone. Corona Virus spreads through human-to-human contact by mouth and nose [10-12]. COVID-19 shows symptomatic and asymptomatic behavior. Many people were affected but due to asymptomatic, they were not able to recognize and affected many other people also and die.

In most of the cases, indications of COVID-19 were dry cough, high body temperature, uncomfortable throat and chills, loss of taste and smell, congestion, nausea, fatigue and joint pain. However, some patients have developed lethal complications like septic shock, severe pneumonia, multiple organ failure and Acute Respiratory Distress Syndrome (ARDS) [13].

Mitigation and preparedness plans are the main basic pillars to getting rid of coronavirus.

Despite extraordinary efforts by the scientific and medical communities to sequence, identify, treat, and prevent COVID-19, some people still have post-COVID symptoms, which can last weeks or months. Different authors have used different terminologies to designate prolonged symptoms following COVID-19 diseases, such as "Long COVID-19," "post-acute COVID-19," "persistent COVID-19 symptoms," "chronic COVID-19," "post-COVID-19 manifestations," "long-term COVID19 effects," "post COVID-19 syndrome," "ongoing COVID-19," "long-term sequelae," or "long-haulers" as synonyms. In recent times, "long-COVID-19," "post-acute sequelae of SARS-CoV-2 infection" and "post-acute COVID-19" terms have been employed [14-16].

COVID-19 is a multi-organ disease and there are reports of permanent and long-term effects of this disease. So it is very essential to understand COVID-19, the post-COVID effect and the status of the administration and individuals dealing with this situation. The present study discusses the knowledge regarding COVID-19, and various post-COVID-19 symptoms reported all over the world regarding professional status, financial status, mental status and status of the government and people dealing with this pandemic situation. In this chapter, we will also go over some of the issues that need to be addressed in the multidisciplinary care of COVID-19 survivors.

2. Post-COVID-19 Symptoms

After COVID-19 onset of symptoms, signs, or abnormal medical issues persist for more than two weeks and the patient does not revert to a normal condition can be considered long-term effects of COVID [17]. Shortness of breath, cognitive impairment, sometimes known as brain fog, and exhaustion are some of the post-COVID-19 syndrome or extended COVID symptoms. These are the most prevalent signs and symptoms. Post-COVID effects impair not just physical health but also emotional, social, and economic well-being. The authors [18] reviewed 34 studies and registered many health problems. The most common problem they have reported were pain (range 4.5%-36%), fatigue (range between 28%-87%), arthralgia (6.0%-27%), declines in physical ability as well as physical role functioning, routine care and everyday activities (range 15%-54% of deceased persons). They were also noted some mental issues like anxiety (range 6.5%-63%), depression (4%-31%) and post-

COVID stress disorder (12.1%-46.9%). Female patients and those admitted to intensive care had higher levels of weariness, pain, anxiety, and sadness.

Pandey et al. [19] surveyed 734 people in India and reported physiological, psychosocial, and economic consequences of the COVID-19 pandemic. During the COVID-19 lockdowns in India, 83% of participants said they had trouble getting healthcare, 17 percent said they had trouble getting medicines, 59 percent said they lost money, 38 percent said they lost jobs, and 28 percent said they ate less fruit and vegetables. Many persons who lost their employment or source of income and had difficulty getting in-patient services reported emotional distress during and after the pandemic. Thus, post COVID-19 effect was not only seen in terms of physical health deterioration but also affected psychological, social and economic conditions.

3. Post-COVID Effect on Physical Health

There have been more than 200 symptoms that people have been experienced post-COVID effect. The most common symptoms were loss of smell and loss of taste, pain in the chest, muscle pains, fever, difficulty in speaking and anxiety or depression. COVID-19 was linked to 55 long-term consequences, according to [20]. These effects were noted in form of headache, joint pain, fatigue, anosmia, ageusia, etc.

Coronavirus is an illness that affects the lungs, as well as other organs such as the heart, kidneys, and mind. Unexpected complications may arise as a result of organ damage that occurs after a Coronavirus infection. Long-term breathing troubles, cardiac issues, extended kidney impairment, stroke, and Guillain-Barre syndrome are all possible long-term health impacts in a small number of people. After recovering from Coronavirus, a small number of adults and children develop the multisystem inflammatory syndrome. A few organs and tissues become severely inflamed as a result of this illness. The effects of post-COVID-19 can be divided into two categories:

- The first category of effect is known as sub-acute or ongoing symptomatic COVID-19,
- The second group includes symptoms and abnormalities that last longer than 12 weeks after the onset of acute COVID-19 [21-23].

Both of the categories of effects may be general (like Nasopharyngeal) or organ-specific (like kidney, liver, lungs specific, etc.). Clinical and scientific proof is advancing on the sub-acute and long-haul impacts of Coronavirus, which can influence different organs [24]. Early studies indicate that SARS-CoV-2 infection has left its mark, with symptoms such as tiredness, dyspnea, chest pain, cognitive difficulties, arthralgia, and a decrease in quality of life [17, 25, 26]. Different side effects were connected with lung infection (like chest distress, cough, diminished pulmonary diffusing limit, pneumonic fibrosis and sleep apnea), cardiovascular (arrhythmias, myocarditis), neurological (dementia, uneasiness, depression, consideration issue, obsessive-compulsive disorders), and others were vague like going bald, tinnitus, and night sweat. Tabacof et al. [27] surveyed 156 pre-vaccinated patients who had been infected with COVID-19 for a median (range) of 351 days (82-457 days). Tiredness, mental fog, and headache were the most common chronic symptoms they described. Tireless indications related to post-intense Coronavirus condition appear to affect physical and intellectual capacity, wellbeing-related personal satisfaction, and participation in society. Table 1 represents organ-specific symptoms of the post-COVID-19 effect.

The 5 most common signs of post-COVID-19 disease were weakness, migraine, attention disorder, going bald, and dyspnea affecting 58%, 44%, 27%, 25% and 24% of people, respectively [20]. Several studies suggested that weakness or tiredness was prominent in females [31, 32]. During or shortly after a COVID-19 infection, some patients, mainly children, develop the multisystem inflammatory syndrome. It’s a condition in which specific body components get inflamed. The multisystem inflammatory syndrome can prompt post-Coronavirus conditions on the off chance that an individual keeps on encountering multi-organ impacts or different manifestations.

Coronavirus infection can occasionally cause post-intensive care syndrome, which refers to health effects that begin while a person is in an emergency room and can last after they return home. Extreme flaws, difficulty with thinking and judgment, and disarray are examples of these effects. Post-traumatic stress disorder (PTSD) refers to the long-term effects of a traumatic experience. This COVID-19 sickness has triggered a global health catastrophe in which countless people are dying, agony is spreading, and lives are being turned upside down [28]. It is not only a physical health crisis but also a social and economic catastrophe, which are both critical for long-term development [29].

Table 1. Organ-specific symptoms in post-COVID-19 effect [28-30]

	Organs	**Symptoms**
1.	Lungs/pulmonary	• Pneumonia associated with coronavirus can cause long-term damage to minute branches alveoli. Symptoms and indications such as dyspnea, impaired exercise capacity, and hypoxia are prevalent. • Obstructive respiratory physiology, and ground-glass opacities and fibrotic alteration on imaging. Long-term breathing issues can be caused by scar tissue in the lungs.
2.	Heart/Cardiovascular	• Even in people who only had a mild COVID impact, irreversible tissue loss of heart muscles is evident in the post-COVID-19 patient. Palpitations, dyspnea, and chest discomfort are common persistent symptoms. • Long-lasting effects can see in form of arrhythmias, myocardial fibrosis or scarring, tachycardia and autonomic dysfunction
3.	Brain	Post-COVID symptoms are reflected in form of neurological symptoms including muscle weakness, tingling or numbness in the hand and legs in young people, and can eventually cause paralysis. It also increases the risk of Alzheimer's and Parkinson's diseases.
4.	Neuropsychiatric	The disease appears in form of fatigue, myalgia, headache, dysautonomia and brain fog. Anxiety, depression, sleep difficulties, and post-traumatic stress disorder (PTSD) have been documented in 30-40% of COVID-19 survivors. Immune dysregulation, medicines, and the psychosocial effects of illness all have a role in the pathophysiology of neuropsychiatric disorders.

	Organs	Symptoms
5.	Blood/Hematologic	In post-acute COVID-19 patients, thromboembolic events were found to be 5% of the time. The duration of the hyperinflammatory state induced by infection with SARS-CoV-2 is unknown. In patients with predisposing risk factors for immobility, more than twice the upper limit of elevated d-dimer, and oedema, direct oral anticoagulants and low-molecular-weight heparin may be used for extended thromboprophylaxis after a risk-benefit discussion.
6.	Kidney	In the Post-COVID case, the patient can develop sudden loss of kidney function, and acute kidney injury.
5.	Endocrine	Endocrine complications may include new or worsened diabetes mellitus management, sub-acute thyroiditis, and demineralization of bones. Patients with newly diagnosed diabetes who do not have traditional type 2 diabetes risk factors, hypothalamic-pituitary-adrenal axis suppression, or hyperthyroidism should have appropriate laboratory testing and be referred to an endocrinologist.
6.	Gastrointestinal and hepatobiliary	COVID-19 has the potential to modify the gut microbiota, including enrichment of opportunistic microbes and decrease of beneficial commensals, even after negative nasopharyngeal swab tests.
7.	Dermatologic	Around 20% of COVID-19 survivors reported hair loss as the most common symptom.
8.	MIS-C *(Multisystem Inflammatory* Syndrome-C)	Criteria for identification: a <21-year-old man with a fever, high inflammatory markers, multiple organ failure, current or recent SARS-CoV-2 infection, and exclusion of alternative diseases Cardiovascular (coronary artery aneurysm) and neurologic (headache, encephalopathy, stroke, and seizure) problems can develop in children aged 7 years and are disproportionate to African, Afro-Caribbean, or Hispanic descent.

4. Post-COVID Effect on Mental Health

According to WHO surveys, the COVID-19 pandemic and Post-COVID conditions have disrupted or stopped vital mental health services in 93 percent of countries globally, despite the rising need. Emerging psychiatric diseases and mental well-being were highlighted as the ninth most common research subject during COVID-19 [33]. The epidemic of COVID-19 is driving up the demand for mental health treatments. Many middle-income countries in Asia face huge monetary difficulties and restricted clinical facilities to keep up with physical and mental prosperity during the pandemic [34]. Isolation, loss of money, and fear are all stimulants that can lead to bad psychological circumstances. Many people may be compelled to increase their drink and drug intake, as well as their sleeplessness and worry. As a result, neurological and mental problems like delirium, anxiety, and stroke can occur. People with mental or neurological impairments are more prone to COVID-19 infection, and they may have a higher risk of catastrophic results, including death. During the international Coronavirus pandemic, comparatively high rates of depression, tension, melancholy, post-traumatic pressure issue, and stress were observed in ordinary people and front-line employees [32, 35].

The authors in [36] studied the impact of COVID-19 on the physical and mental health of seven middle-income Asian countries. According to them, male sexual orientation, sticking with children or more than six people in the same household, work, trust in doctors, a high projected possibility of survival, and investing less energy in health data are all protective factors. The authors in [37] examined the impact of COVID-19 on mental health and quality of life in the Middle East and North Africa region's general population. The COVID-19 epidemic made 45 percent to 62 percent of people feel terrified, nervous, or powerless. Furthermore, over 40% of respondents said employment and financial concerns had worsened their stress levels. Advanced levels of psychological symptoms were found among females, 26-35 years old persons and lower educational group people. The COVID-19 infection was not only linked to a modest psychological impact, but it also promoted family support and mental health awareness among adults in the study area. According to certain studies, social distance and security measures have an impact on people's relationships and their perceptions of empathy for others. Telepsychology and technology equipment, in this context, play critical roles in reducing the pandemic's detrimental impacts. These tools offer advantages that could improve online psychiatric treatment for patients, such as the ability to meet from home or at work, saving money and time, and maintaining a therapist-patient relationship [38].

Another emerging concern of the global pandemic is seen in form of post-COVID stress disorder. This disorder is most common in people or groups who have experienced serious coronavirus disease and possibly death; individuals who, as family members or health care workers, have witnessed others' suffering and death; individuals who have learned about the death or risk of death of a family member or friend due to the COVID-19; and individuals who have experienced extreme exposure to aversive details such as journalists, doctors, first responders, and others [36, 39]. Emotional costs, such as social isolation, unemployment and economic losses, and working from home while caring for children and other family members, are additional post-COVID pressures.

5. Post-COVID Effect on the Social Condition

The COVID-19 pandemic has engulfed the entire world, affecting all segments of the population, and is particularly harmful to members of those categories of people who are in the most vulnerable situations, such as people living in poverty, and the elderly people, disabled people, youth, and indigenous peoples.

Initially, it was thought that impoverished people were bearing the brunt of the infection's health and financial consequences. Homeless people, people without access to running water, travelers, refugees, and other dislocated people, for example, continue to suffer from the effects of the Coronavirus and its aftermath - whether as a result of limited mobility, or fewer employment opportunities, or increased xenophobia, for example [40]. If not properly addressed through policy the social crises created by this pandemic may likewise build imbalance, avoidance, segregation and worldwide joblessness in the medium and long term. Extensive, widespread social security frameworks, when set up, assume a much more durable role in protecting workforces and in reducing the prevalence of poverty since they act as automatic stabilizers [41]. That is, they provide basic income safety at all times, accordingly improving peoples' ability to manage and overcome losses.

In India, during the pandemic lockdown, the movement of the entire 1.3 billion population was limited and the pandemic has left a visible impression across all sectors, the impact on families living in poor and unsatisfactory conditions, women, and children has been more. Gender-based violence, lack of security, money, and health have added to the existing misery of families living in poor and substandard conditions. Social distancing is by all accounts hitting individuals significantly more than the panic of the dangerous infection. People have developed to be social animals and are bound to live in intuitive gatherings. Being disconnected from family, companions, and associates can be unbalancing and awful for the vast majority and can bring about short or even long-term mental and actual medical issues. An expansion in the degrees of nervousness, aggression, despair, amnesia, and hallucinations are all symptoms of schizophrenia. Individuals are getting exceptionally anxious and disturbed despite online media availability. Starvation, suicides, tiredness, road and train accidents, police abuse, and denial of timely medical assistance were all mentioned as causes of death during the lockdown. Lockdown in India has also resulted in positive effects; it has provided a perfect opportunity for those who rarely have time to spend with their families. Individuals were encouraged to work from home and spend more time with their parents, spouses, and children as a result of the lockdown.

6. Post-COVID Effect on the Economic Condition

As a result of the post-COVID-19 effect, public health is in danger [42] and has a significant impact on the global economy. During the pandemic, the global lockdown had several consequences, including job losses, business closures, and stock market crashes. In 2020, the global economy shrank as a result of COVID. Due to two key elements, the COVID-19 pandemic harms economic growth, first, the global epidemic's exponential growth contributed directly to a great deal of uncertainty about financial and capital market instability. Second, countries have imposed stringent restrictions on human mobility and transportation, resulting in a major reduction in economic activity. This has put stress on consumers and producers both. As a result, post-COVID-19 has seen in terms of an economic downturn. The safety procedures taken to save lives are seriously influencing the economies of the entire world. Protection measure like lockdown has constrained individuals to work from home wherever possible. Closure of workstations has disturbed stock chains and brought down usefulness. Other economic protection measures, such as travel bans and the prohibition of sporting events and other large gatherings, have had an impact. In addition, in many parts of the world, measures such as discouraging the use of public transportation and public spaces, such as restaurants, shopping malls, and public attractions, have been implemented. The situation was particularly worse in hospitality-related industries and the worldwide travel industry, with airlines, cruise lines, casinos, and hotels reporting a 90 percent drop in commercial activity [43]. Businesses that rely on social contacts, such as entertainment and tourism, have been hit hard, and millions of people have lost their employment as a result. People are spending less as a result of lower personal income and uncertainty, leading to more business closures and job losses [44]. Gross Domestic Production (GDP) is expected to decrease by 3% in 2020, according to forecasts [45]. Most advanced economies, including the United States, Japan, the United Kingdom, Germany, France, Italy, and Spain, will decrease by 5.9%, 5.4 percent, 6.5 percent, 7.0 percent, 7.2 percent, 9.1 percent, and 8.0 percent in 2020, respectively. According to the April 2020 World Economic Outlook, GDP levels in advanced, emerging markets, and developing nations will stay below pre-virus levels by the end of 2021 [46].

In India, too, the post-Coronavirus epidemic has been unpleasant in terms of economic activity and human lives lost. Indeed, even before the Coronavirus, India's economy was in a slump, with unprecedented levels of unemployment and poverty [47]. After March 2020, economic crises impacted all areas of the Indian economy. An expected 140 million individuals lost their work during this pandemic, while compensations were cut for some others [48]. Approximately 45 percent of households in the United States have reported a decrease in income from the previous year. As a result, when the epidemic struck in March 2020, India's ability to deal with a new catastrophe was limited. Farmers faced disrupted supply chains, a shortage of market outlets, low demand, and falling output prices in the agriculture sector. Micro and small businesses were the most heavily impacted segments in the industrial sector.

Approximately 15 million people were laid off as a result of the crisis. In the industry, Miniature and little endeavors were the most heavily impacted. Ramakumar [49] predicted that India's GDP growth rate in 2020-2021 might range from 4.3 percent to 15 percent using an input-output (I-O) paradigm. According to a report, Larsen & Toubro, Bharat Forge, UltraTech Cement, Grasim Industries, Aditya Birla Group, BHEL, and Tata Motors are among the major Indian companies that have temporarily halted or restricted operations. As funding has dried up, new start-ups have been harmed. Many commodities businesses in India have drastically reduced their operations and are concentrating on the necessities. Stock markets suffered their greatest losses in history on March 23, 2020.

Conclusion

COVID-19 is a multi-organ disorder, and there have been instances of long-term repercussions. This can be acute or chronic. Post-COVID symptoms have not only been seen in terms of physical health injury, but it also affects the social, mental and economic condition of an individual. All types of rehabilitation, physical, pulmonary, cognitive, and mental health support are important tools to be employed in the post-acute care of COVID-19 [50-52]. These involvements go hand-in-hand and can benefit to increase the quality of life and function for many Coronavirus patients. So, multidisciplinary care for COVID-19 survivors is the basic need for an hour.

References

[1] Jain, S., Kumar, M., Sindhwani, N., & Singh, P. (2021, September). SARS-Cov-2 detection using Deep Learning Techniques on the basis of Clinical Reports. In: *2021 9th International Conference on Reliability, Infocom Technologies and Optimization (Trends and Future Directions) (ICRITO)* (pp. 1-5). IEEE.

[2] Jain, S., Sindhwani, N., Anand, R., & Kannan, R. (2022). COVID Detection Using Chest X-Ray and Transfer Learning. In: *International Conference on Intelligent Systems Design and Applications* (pp. 933-943). Springer, Cham.

[3] Khongsai, L., Anal, T. S., As, R., Shah, M., & Pandey, D. (2021). Combating the spread of COVID-19 through community participation. *Global Social Welfare*, 8(2), 127-132.

[4] Pandey, D., Ogunmola, G. A., Enbeyle, W., Abdullahi, M., Pandey, B. K., & Pramanik, S. (2021). COVID-19: A framework for effective delivering of online classes during lockdown. *Human Arenas*, 1-15.

[5] Fehr, A. R., & Perlman, S. (2015). Coronaviruses: an Overview of Their Replication and Pathogenesis, *Methods in Molecular Biology,* 1282, 1-23.

[6] Lu, H., Stratton, C. W., & Tang, Y. W. (2020). Outbreak of pneumonia of unknown etiology in Wuhan, China: The mystery and the miracle. *Journal of Medical Virology*, 92(4), 401-402. https://doi.org/10.1002/jmv.25678.

[7] Sohrabi, C., Alsafi, Z., O'Neill, N., Khan, M., Kerwan, A., Al-Jabir, A., Iosifidis, C., & Agha, R. (2020). World Health Organization declares global emergency: A review of the 2019 novel coronavirus (COVID-19). *International Journal of Surgery* (London, England), 76, 71-76. https://doi.org/10.1016/j.ijsu.2020.02.034.

[8] Dong, E., Du, H. & Gardner L. (2020). An interactive web-based dashboard to track COVID-19 in real-time. *Lancet Infectious Disease* 20(5), 533-534.

[9] World Health Organization, WHO, (2020). Director-General's Remarks at the Media Briefing on 2019-nCoV on 11 February 2020. https://www.who.int/dg/

speeches/detail/who-director-general-s-remarks-at-the-media-briefing-on-2019-ncov-on-11-february-2020.

[10] Anand, R., Sindhwani, N., & Saini, A. (2021). Emerging Technologies for COVID-19. *Enabling Healthcare 4.0 for Pandemics: A Roadmap Using AI, Machine Learning, IoT and Cognitive Technologies*, 163-188.

[11] Pandey, D., Islam, T., & Malik, M. A. (2021). Novel coronavirus disease (SARS-COV-2): an overview. *Asian Journal of Advances in Medical Science*, 8-12.

[12] Meivel, S., Sindhwani, N., Anand, R., Pandey, D., Alnuaim, A. A., Altheneyan, A. S., & Lelisho, M. E. (2022). Mask Detection and Social Distance Identification Using Internet of Things and Faster R-CNN Algorithm. *Computational Intelligence and Neuroscience, 2022.*

[13] Chen, N., Zhou, M., Dong, X., Qu, J., Gong, F., Han, Y., Qiu, Y., Wang, J., Liu, Y., Wei, Y., Xia, J., Yu, T., Zhang, X., & Zhang, L. (2020). Epidemiological and clinical characteristics of 99 cases of 2019 novel coronavirus pneumonia in Wuhan, China: a descriptive study. *Lancet* (London, England), 395(10223), 507-513. https://doi.org/10.1016/S0140-6736(20)30211-7.

[14] Rubin, R. (2020). As their numbers grow, COVID-19 "long haulers" stump experts. *JAMA* 324(14), 1381-1383. https://doi.org/10.1001/jama.2020.17709 (2020).

[15] Pandey, D., Babu, S. Z. D., Noibi, T. O., Pandey, B. K., Dey, P. T. D., Kassaw, C., & Canete, J. J. O. (2020). A study on novel covid-19 epidemic threat for universal physical situation in 2020. *Asian Journal of Advances in Research*, 27-33.

[16] Parthiban, K., Pandey, D., & Pandey, B. K. (2021). Impact of SARS-CoV-2 in online education, predicting and contrasting mental stress of young students: a machine learning approach. *Augmented Human Research*, 6(1), 1-7.

[17] Tenforde, M. W., Kim, S. S., Lindsell, C. J., Billig Rose, E., Shapiro, N. I., Files, D. C., Gibbs, K. W., Erickson, H. L., Steingrub, J. S., Smithline, H. A., Gong, M. N., Aboodi, M. S., Exline, M. C., Henning, D. J., Wilson, J. G., Khan, A., Qadir, N., Brown, S. M., Peltan, I. D., Rice, T. W., IVY Network Investigators (2020). Symptom Duration and Risk Factors for Delayed Return to Usual Health Among Outpatients with COVID-19 in a Multistate Health Care Systems Network - United States, March-June 2020. *MMWR. Morbidity and Mortality Weekly Report*, 69(30), 993-998. https://doi.org/10.15585/mmwr.mm6930e1.

[18] Shanbehzadeh, S., Tavahomi, M., Zanjari, N., Ebrahimi-Takamjani, I., & Amiri-arimi, S. (2021). Physical and mental health complications post-COVID-19: Scoping review. *Journal of Psychosomatic Research*, 147, 110525. https://doi.org/10.1016/j.jpsychores.2021.110525.

[19] Singh et al. (2021). Health, psychosocial, and economic impacts of the COVID-19 pandemic on people with chronic conditions in India: a mixed methods study. *BMC Public Health* 21:685. https://doi.org/10.1186/s12889-021-10708-w.

[20] Lopez-Leon, S., Wegman-Ostrosky, T., Perelman, C., Sepulveda, R., Rebolledo, P. A., Cuapio, A., & Villapol, S. (2021). More than 50 Long-term effects of COVID-19: a systematic review and meta-analysis. *medRxiv: the preprint server for health sciences*, 2021.01.27.21250617. https://doi.org/10.1101/2021.01.27.21250617.

[21] Horvath, L., Lim, J., Taylor, J. W., Saief, T., Stuart, R., Rimmer, J., & Michael, P. (2021). Smell and taste loss in COVID-19 patients: assessment outcomes in a Victorian population. *Acta Oto-Laryngologica*, 141(3), 299-302.

[22] Kamal, M., Abo Omirah, M., Hussein, A., & Saeed, H. (2021). Assessment and characterisation of post-COVID-19 manifestations. *International Journal of Clinical Practice*, 75(3), e13746. https://doi.org/10.1111/ijcp.13746.

[23] Mandal, S. (2020). "Long-COVID": A cross-sectional study of persisting symptoms, biomarker and imaging abnormalities following hospitalisation for COVID-19. *Thorax* https://doi.org/10.1136/thoraxjnl-2020-215818.

[24] Gupta, A. (2020). Extrapulmonary manifestations of COVID-19. *Nature Medicine,* 26(7), 1017-1032.

[25] Carfi, A., Bernabei, R., Landi, F., & Gemelli against COVID-19 Post-Acute Care Study Group (2020). Persistent Symptoms in Patients After Acute COVID-19. *JAMA*, 324(6), 603-605. https://doi.org/10.1001/jama.2020.12603.

[26] Huang, C. (2021). 6-month consequences of COVID-19 in patients discharged from hospital: a cohort study. *Lancet* 397(10270), 220-232.

[27] Tabacof L., Tosto-Mancuso J., Wood J., Cortes M., Kontorovich A., McCarty D., Rizk D., Rozanski G., Breyman E., Nasr L., Kellner C., Herrea E., & Putrino D. (2022). Post-acute COVID-19 Syndrome Negatively Impacts Physical Function, Cognitive Function, Health-Related Quality of Life, and Participation. *American Journal of Physical Medicine and Rehabilitation*. 101(1): 48-52.

[28] Nicola, M. (2020). The socio-economic implications of the coronavirus pandemic (COVID-19): a review. *International Journal of Surgery*, 78, 185-193.

[29] Pirouz, B. (2020). Investigating a serious challenge in the sustainable development process: analysis of confirmed cases of COVID-19 (new type of coronavirus) through a binary classification using artificial intelligence and regression analysis. *Sustainability* 12 (6), 427.

[30] Nalbandian (2021). Post-acute COVID-19 syndrome. *Nature Medicine*, 27: 601-615.

[31] Townsend, L., Dowds, J., O'Brien, K., Sheill, G., Dyer, A. H., O'Kelly, B., Hynes, J. P., Mooney, A., Dunne, J., Ni Cheallaigh, C., O'Farrelly, C., Bourke, N. M., Conlon, N., Martin-Loeches, I., Bergin, C., Nadarajan, P., & Bannan, C. (2021). Persistent Poor Health after COVID-19 Is Not Associated with Respiratory Complications or Initial Disease Severity. *Annals of the American Thoracic Society*, 18(6), 997-1003. https://doi.org/10.1513/AnnalsATS.202009-1175OC.

[32] Xiong, J., Lipsitz, O., Nasri, F., Lui, L.M., Gill, H., Phan, L., (2020). Impact of COVID-19 pandemic on mental health in the general population: A systematic review. *Journal of Affective Disorders,* 277, 54-64.

[33] Tran, B. X., Ha, G. H., Nguyen, L. H., Vu, G. T., Hoang, M. T., Le, H. T., Latkin, C. A., Ho, C., & Ho, R. (2020). Studies of Novel Coronavirus Disease 19 (COVID-19) Pandemic: A Global Analysis of Literature. *International Journal of Environmental Research and Public Health*, 17(11), 4095. https://doi.org/10.3390/ijerph17114095.

[34] Fitzgerald, D. A, & Wong, G. W. (2020). COVID-19: A tale of two pandemics across the Asia Pacific region. *Paediatric Respiratory Reviews*. 35, 75-80. https://doi.org/10.1016/j.prrv.2020.06.018 PMID: 32768308.

[35] Khan, Y. H., Mallhi, T. H., Alotaibi, N. H., & Alzarea, A. I. (2020). Work related stress factors among healthcare professionals during COVID-19 pandemic; a call for immediate action. *Hospital Practice (1995)*, 48(5), 244-245. https://doi.org/10.1080/21548331.2020.1793519.

[36] Wang, C., Tee, M., Roy, A. E., Fardin, M. A., Srichokchatchawan, W., Habib, H. A., (2021). The impact of COVID-19 pandemic on physical and mental health of Asians: A study of seven middle income countries in Asia. *PLOS ONE* 16(2), e0246824. https://doi.org/10.1371/journal.pone.0246824.

[37] Al Dhaheri, A. S., Bataineh, M. F., Mohamad, M. N., Ajab, A., Al Marzouqi, A., Jarrar, A. H., (2021) Impact of COVID-19 on mental health and quality of life: Is there any effect? A cross-sectional study of the MENA region. *PLOS ONE 16(3): e0249107*. https://doi.org/10.1371/journal.pone.0249107.

[38] Saladino, V., Algeri, D., & Auriemma, V. (2020). The Psychological and Social Impact of COVID-19: New Perspectives of Well-Being. *Frontiers in Psychology*, 11. https://doi.org/10.3389/fpsyg.2020.577684.

[39] Bo, H. X., Li, W., Yang, Y. (2021). Posttraumatic stress symptoms and attitude toward crisis mental health services among clinically stable patients with COVID-19 in China. *Psychoogiel Medicale* 27, 1-2.

[40] UN DESA. (2021). *The UN Department of Economic and Social Affairs* https://www.un.org/coronavirus.

[41] Mofijur (2021). Impact of COVID-19 on the social, economic, environmental and energy domains: Lessons learnt from a global pandemic. *Sustainable Production and Consumption* 26, 343-359.

[42] Chinazzi, M. (2020). The effect of travel restrictions on the spread of the 2019 novel coronavirus (COVID-19) outbreak. *Science* 368 (64 89), 395-40.

[43] Fernandes, N. (2020). Economic effects of coronavirus outbreak (COVID-19) on the world economy (March 22, 2020). *IESE Business School Working Paper No. WP-1240-E*, Available at SSRN: https://ssrn.com/abstract=3557504 or http://dx.doi.org/10.2139/ssrn.3557504.

[44] Ghosh, A., Nundy, S., & Mallick, T. K. (2020). How India is dealing with COVID-19 pandemic. *Sensor International* 1, 100021, https://doi.org/10.1016/j.sintl.2020.100021.

[45] International Monetary Fund (IMO) 2021. "Policy Responses to COVID19." Archived from the original on 20 May 2021.

[46] WHO (World health Organization). (2021). Coronavirus disease (COVID-19) pandemic. https://www.who.int/COVID-19.

[47] Parvathamma, G. L. (2021). Socio-Economic Impact of the first wave of COVID-19 Pandemic on Indian Economy-A Conceptual study. *Journal of Economics and Finance*, 2(4), III: 1-6.

[48] Pandey, S., Gupta, S., Chhajed, S. (2020). 3Ps-Path, Pace and Pattern of Recovery from COVID-19 for different businesses in Indian market. *International Journal of*

Engineering Research & Technology (IJERT), 9(7): http://dx.doi.org/10.17577/IJERTV9IS070545.

[49] Ramakumar, R. (2020). Agriculture and the COVID-19 Pandemic: An Analysis with special reference to India. *Review of Agrarian Studies*, 10(1), pp. 72-110.

[50] Singh, H., Rehman, T. B., Gangadhar, C., Anand, R., Sindhwani, N., & Babu, M. (2021). Accuracy detection of coronary artery disease using machine learning algorithms. *Applied Nanoscience*, 1-7.

[51] Juneja, S., Juneja, A., & Anand, R. (2020). Healthcare 4.0-Digitizing Healthcare Using Big Data for Performance Improvisation. *Journal of Computational and Theoretical Nanoscience*, 17(9-10), 4408-4410.

[52] Shukla, R., Dubey, G., Malik, P., Sindhwani, N., Anand, R., Dahiya, A., & Yadav, V. (2021). Detecting crop health using machine learning techniques in smart agriculture system. *Journal of Scientific and Industrial Research (JSIR)*, 80(08), 699-706.

Chapter 8

Detection and Analysis of Mental Stress in Online Learning Using IoT Devices

Veeresh Rampur[1,*], Y. Prasanna Kumar[2,†], Santanu Das[3,‡], Shahanawaj Ahamad[4,§], G. Nageswara Rao[5,‖], Sushma Jaiswal[6,¶] and Ankur Gupta[7,#]

[1]Department of Electronics, Government First Grade College, Bidar, Karnataka, India
[2]Department of Mining Engineering, BuleHora University, BuleHora, Ethiopia
[3]Department of Biotechnology, Seshadripuram First Grade College, Yelahanka New Town, Bangalore, India
[4]College of Computer Science and Engineering, University of Hail, Hail City, Saudi Arabia
[5]Department of EEE, Lakireddy Bali Reddy College of Engineering, Mylavaram, Andhra Pradesh, India
[6]Department of Computer Science & Information Technology, Guru GhasidasVishwavidyalaya, Bilaspur, Chhattisgarh, India
[7]Department of Computer Science and Engineering, Vaish College of Engineering, Rohtak, Haryana, India

[*] Author's Email: vvrampur@gmail.com.
[†] Author's Email: prasannaky@gmail.com.
[‡] Author's Email: santanudas1stdec@gmail.com.
[§] Author's Email: drshahwj@gmail.com.
[‖] Author's Email: gnrgudipudi@gmail.com.
[¶] Author's Email: jaiswal1302@gmail.com.
[#] Corresponding Author's Email: ankurdujana@gmail.com.

In: An Interdisciplinary Approach in the Post-COVID-19 Pandemic Era
Editors: D. Pandey, B. Kumar Pandey, N. Sindhwani et al.
ISBN: 979-8-88697-052-4

Abstract

The effects of stress are causing rigorous damage to the mental as well as physical state of humans. It is very difficult to identify whether a person is in stress. The person may look healthier physically but may not be in a state of good health due to the stress within the body. Their mental stability also gets affected and may further lead to chronic ailments due to persistent stress. However, it is very essential to monitor stress levels regularly which help in diagnosis of any abnormalities in the body that may lead to chronic illness in future. The Wireless networks based on IoT provides wide range of opportunities to monitor stress levels regularly and transmit the information to the concerned for immediate action. A model is designed and developed to detect the stress levels using various sensors like heartbeat rate, BP, body temperature and concentration of CO_2 gas. Further based on the values of these sensors, the levels of stress are calculated and the information is transmitted using IoT.

Keywords: Internet of Things, online learning, mental stress

1. Introduction

Internet of Things (IoT) is no longer a mystery. Until recently, it was a technology that was silently reshaping our future. The Internet of Things was created as a consequence of human inventiveness and a desire to lead a more pleasant and interconnected existence [1-4]. We decided to make electronics smart and take care of the things that suck up time. This step has produced good results. When people and objects are connected and interact with one other, it is known as the IoT. IoT has role to play in the healthcare industry because of its human component, which may influence people's health habits for the better [5]. A variety of technology solutions have been created to improve healthcare and execute preventative activities, notably in the field of metal stress [6]. Provision of ubiquitous healthcare solutions which give healthcare services at anytime anyplace has grown more attractive today owing to the focus on healthcare awareness and also the expansion of mobile wireless technology [7].

Stress has become a serious issue in today's culture. Almost everyone, including students and employers, is working feverishly to fulfill their deadlines, goals, and other deadlines. Often, individuals are aware that they are under a lot of stress at work, but they don't pay attention to their physical health [8, 9]. They may also neglect to take their medications on time, which may have dangerous side effects or even result in death. Unchecked sensor values, such as heart rate and blood pressure, might be worrisome. Heart attacks and fatalities may be prevented if the proper treatment is taken at the right time. This means that the development of stress detection and health monitoring technologies [10] that can assist individuals comprehend their mental and physical states are very necessary. Biomedical applications of wireless technology have grown significantly in recent years, with the goal of improving patient care among them. Continuous monitoring devices [11] are being produced in large numbers [12]. There are numerous current systems that capture and save data on a generic storage server that is accessible only to the staff and physicians [13]. Heart rate, blood pressure, temperature, humidity, and respiration will all be tracked by a model in the proposed system. Data from these sensors will be transferred to the server through a Wi-Fi module [14]. The GSM module may be used to send a message to a concerned person or doctor [15]. The device uses less power and has been developed to accurately measure stress levels.

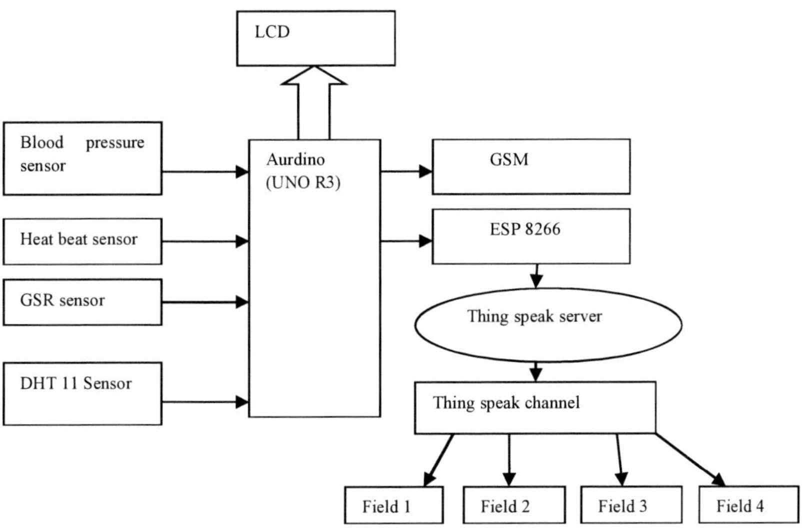

Figure 1. Block diagram of proposed IoT based stress detection system.

IoT-based stress detection is shown in Figure 1 as a block diagram with various sensors for temperature, heart rate [16], blood pressure[17], gas values being sensed by the sensors [18, 19]. Data is processed and stored and, if a microcontroller detects an abnormal value for any parameter, an SMS alert [20] is sent to a mobile device [21] via a GSM module to the person or doctor [22]. With the help of the Internet of Things [23], healthcare providers can keep a closer eye on their patients and establish more frequent, proactive connections with them [24]. Physicians may use IoT device data to establish the optimal treatment plan for each patient [25] and achieve the desired results [26].

1.1. Internet of Things

IoT was established to enable intelligent devices to connect with each other as the next great disruptive technology sector. It integrates digital and physical transactions by integrating various technologies from various applications. Smart homes are one of the most widely utilized IoT applications, including sensors, hardware and software systems, and computer platforms. Many cutting-edge gadgets with some degree of artificial intelligence are now referred to as "smart," not only those that may be found in the common household. The Internet of Things (IoT) may help increase system automation, analysis, and integration. They broaden and fine-tune these fields. Both existing and cutting-edge technology are used to create an Internet of Things (IoT). Software breakthroughs, hardware price reductions and current attitudes on technology all contribute to IoT's growth [18]. As a consequence of its new and sophisticated aspects, the supply of products, commodities, and services has undergone a major transformation. An increasing number of natural resources are being depleted at an alarming pace, which necessitates Internet of Things (IoT) monitoring of the environment. As part of normal environmental monitoring procedures, scientists collect, inspect, and analyses samples. There are two ways to go about this. It is done manually in a laboratory in the first instance. It is possible to count the amount of pollutants in a sample using the instrumental approach. Instrumentation eliminates the need for tedious data entering and analysis. In order to carry out sedimentation, isolation, and other treatments on the sample, the sample must first be pre-treated manually. The topics of waste management, air pollution, and severe weather are often brought up while addressing IoT-enabled environmental monitoring.

1.2. Online Learning (E-Learning)

Student loyalty and devotion are fostered in the traditional educational system by having students and professors interact in actual classrooms on campus. However, the use of ICTs in many industries and sectors, including high school, has evolved as a result of this advancement. With the increasing use of ICT applications, the use of e-learning in high schools is becoming more commonplace. An innovative approach to education services may be provided through the use of electronic forms of information to enhance the knowledge and abilities of learners via the use of e-learning [27, 28]. To put it another way, e-learning makes use of contemporary ICTs & Internet-based computers to provide educational content. E-learning has a number of benefits for both institutions and students. First and foremost, e-learning saves money for universities by removing the need to spend in physical education and learning infrastructure. It also helps to digitizing academic institutions and creating a knowledge-based digital society where studying and sharing information can be accomplished swiftly and conveniently over the Internet from any location [29, 30].

There are three reasons why e-learning is so important in today's educational landscape. Local students do not have to go to a foreign country in order to benefit from international institutions' full range of academic services while taking participation in common training programs that take place inside the country's boundaries. E-learning provides students with an additional technique of learning in addition to traditional methods. As long as a computer or mobile device is connected to the Internet and the University's e-learning system, e-learning may be done from home, at work, or anywhere else. This is a huge aid for those who are juggling school and job at the same time. A final benefit of online education is that it frees students from the constraints of traditional classroom attendance.

1.3. Mental Stress

Anxiety and anguish caused by a person's perception of events in their external or internal surroundings, a kind of stress. Physiological reactions are often accompanied by mental tension. Psychologists define stress as a state of mental or emotional exhaustion or tension. Psychological discomfort takes the form of stress. Stress may have a positive effect on sports performance, motivation, and response to the environment in small doses. Working too much or not doing enough in your life. Emotional or bodily strain may lead to a sensation of stress. It's your body's response to a problem or a demand that causes you stress. However, long-term stress can be detrimental.

1.4. An IoT System for Stress Management

A sensor-based monitoring system typically includes the following components: sensors that gather data, displays that show measured values, data exchange, and devices that are wearable, precise, long lasting, and trustworthy [20]. Three components make up the IoT system architecture for stress measurement: a wearable device for monitoring vital indicators, a mobile health app with relaxing material, and a cloud platform. Additional services for connecting components, hosts, and users have been created by our team as well. A heart rate sensor, a microcontroller to interpret the data, and a wireless Internet connection are all necessary for the wearable device to function properly [21]. The relaxing app may be downloaded to any Android phone or tablet with an active Internet connection for the second component. Data from the wearable system's sensors is sent to the cloud, where it is analyzed and the user's history is tracked [22, 23].

1.5. Role of Mental Stress with IoT

Various IoT and wearable technologies can assess a user's vital signs and detect whether or not they are under stress [18]. Commercial IoT systems do not have open APIs, making them difficult to link with other e-education services since they cannot be programmed and modified. We've opted for an open architectural approach to make this connection feasible. IoT stress management may be integrated into an educational environment because to its open design. Students' anxiety should be reduced, as should their stress levels in the future, thanks in part to a mobile health application that provides relaxing material [25].

1.6. Role of Mental Stress in Online Learning

Anxiety and arousal may result from stressful events. Aroused people's bodies are put under a lot of stress. Because of the sympathetic nervous system's involvement in human behaviors such as sweating, elevated heart rate, and tightened muscles, stress triggers a variety of physiological responses. Despite the fact that stress aids performance, excessive or repetitive exposure to stress may have detrimental consequences on health. It is well acknowledged that stress is one of the most pressing issues in healthcare, and that it has a significant influence on people's well-being. Recurring stress symptoms over time in any group may serve as an early warning sign for various health problems [26].

1.7. Hardware Description

1.7.1. Arduino

The ATmega328 comprises 14 pins on the Arduino UNO microcontroller. The Arduino board does not have the ability to run code without the aid of an external power source. Integrated software on the Arduino board detects if the board is functioning properly. With sensors attached, the Arduino board requires just 5 volts of power to communicate with each other.

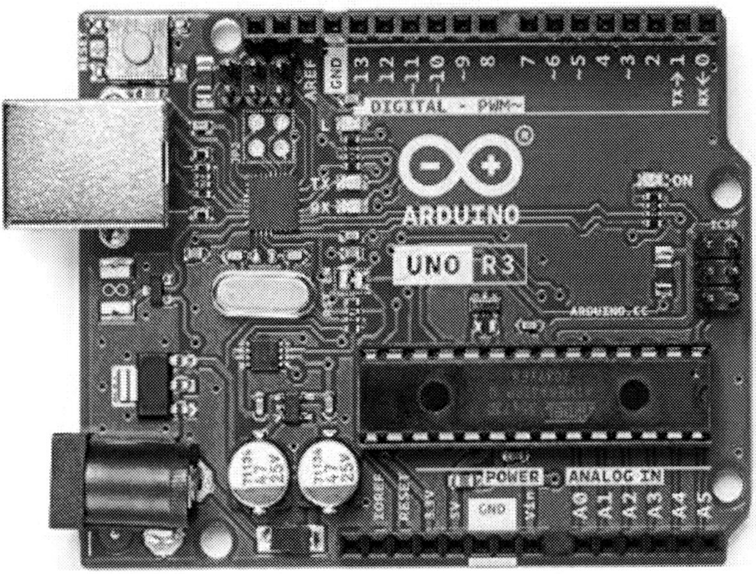

Figure 2. Arduino UNO R3 [31].

1.7.2. Blood Pressure Sensor

Sensor (number) monitors and displays the values of Blood Pressure and Pulse rate and sends the measured values to other devices through serial connection. – – Systolic, Diastolic, and Pulse Rate are shown on the display (as shown in Figure 3). It resembles a wristwatch that does away with the need for pumping.

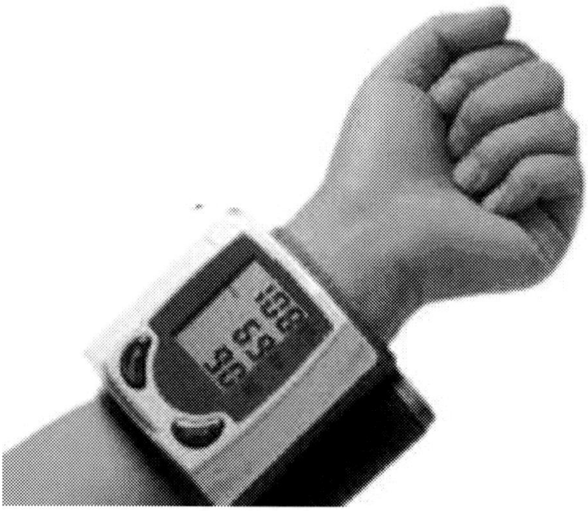

Figure 3. BP Sensors [32].

1.7.2.1. Features of the BP Sensor

- Smart automatic compression and decompression
- 60 memory groups for storing the value measurements
- Power saving mode if it will sit idle for more than 3 minutes
- Large display screen
- Power specification: +5V
- Communicates with other devices serially
- Allows compression and decompress

1.7.2.2. Specifications of the Blood Pressure Sensor

It operates at 5V, 200ma voltage regulated and each reading consists of 15 bytes at 9000 baud rate. The output reading is 8-bit value in ASCII format fixed digits from 000 to 255.

Table 1. Standard classification chart of blood pressure (values) for adults

Name of the Stage	Systolic (mm Hg)	Diastolic (mm Hg)
Hypotension	<90	<60
Desired	90-119	60-79
Pre Hypertension	120—139	80-89-
Stage 1 Hypertension	140-159	90-99
Stage 2 Hypertension	160-179	90-99
Hypertensive Crisis	>=180	>=110

1.7.3. GSM (Global System for Mobile Communication)

RS32 is a serial port on a GSM module (shown in Figure 4) that can communicate with an external fringe. To access computerized data, it uses a basic transporter flag and then demodulates to comprehend the clear data. Open and digital, GSM is used to carry mobile voice and data services. Using two distinct streams of client data, each in its own specific time slot, a GSM digitalizes and reduces the data before sending it down the channel.

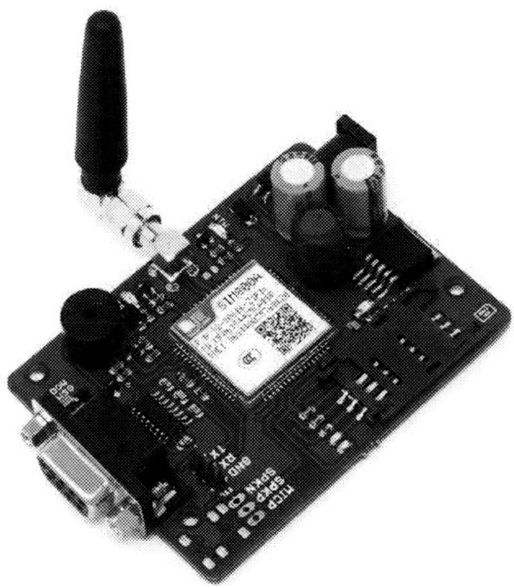

Figure 4. GSM module [33].

1.7.4. DHT11 Temperature

It measures both temperature and humidity with the DHT11 (as shown in Figure 5). It monitors the surrounding temperature and humidity. Any microcontroller may be interfaced with this sensor, and the data can be sent to the microcontroller. Humidity may be measured in range of 20 percent to 90 percent with accuracy of 0.1°C & 0.1 percent, respectively.

Figure 5. DHT11 (LM35) Sensor [34].

1.7.5. NodeMCU: ESP8266

The ESP8266 (as shown in Figure 6) is a Wi-Fi module embedded into the SOC with a transfer control protocol and an internet convention stack that may be used by the controller to connect to the Internet. The ESP8266 Wi-Fi module is used to transmit data from the Arduino to Thingspeak server wirelessly.

Figure 6. NodeMCU: ESP8266 [35]

1.7.5.1. Software Requirements

Programmers of Aurduino may benefit from the Arduino IDE (Integrated Development Environment) software, which offers full tools for software development. The application is written in C and then uploaded to the Arduino using the C programming language [17].

2. Literature Review

To see &analyses ECG waveforms from wearable ECG devices in real time through a wireless sensor network, Pei - Cheng Hii [7] suggested a mobile monitoring terminal in 2011(WSN). Using WSN in healthcare allows for the substitution of wired technology, bringing healthcare out of a fixed geographic setting and into the cloud. Barcode decoding on mobile phones is used as an extension of the monitoring programme to check and support outpatients in drug administration process, giving more complete healthcare service.

HealthGear, real-time wearable device for monitoring, viewing, and interpreting physiological information, was introduced by Health Gare [8] in 200 Using Bluetooth, HealthGear connects noninvasive physiological sensors to a mobile phone that saves, transmits and analyses the data. The data is then presented to the user in an understandable manner. HealthGear is a system that monitors a user's vital signs, and blood oxygen and heart rate, while they sleep. Two distinct sleep apnea detection algorithms are also described, as well as the entire system's performance in a sleep study with 20 participants.

A technique called TensorBeat was suggested by Iman [10] in 2017 that uses CSI phase difference data to automatically predict the breathing rates of many people using Wi-Fi devices. The tensor decomposition approach may be used to handle phase difference data from the CSI. In order to build CSI tensors, the proposed TensorBeat method first gathers CSI phase difference data between antenna pairs at Wi-Fi receiver. Breathing signals are then extracted using canonical polyadic decomposition (CP). Multiple people's breathing rates may be estimated using a peak detection approach in conjunction with a reliable signal matching system. In our experiment, we found that TensorBeat can accurately monitor the breathing rate of multiple people in a variety of environments.

A HiCH for IoT-based health monitoring systems was proposed by Arman [9] in 2017 as a solution to these problems. (1) A new computational architecture that allows for hierarchical division & execution of machine learning-based data analytics, (2) a management approach for closed-loop self-adaptation depending on the patient's state are the two main components of the proposed system. For hospital IoT systems, he developed a unique approach to management. HiCH's ability to identify arrhythmias in patients with cardiovascular disease is shown in a complete performance assessment (CVDs).

A system based on air-pressure sensors and air bladders was suggested by Junk [12] in 2016. The change in air pressure in an air-bladder in contact with targeted muscle is used to detect muscular activation (s). SNR and reaction time make the proposed sensing approach advantageous for mobile devices because the change in air pressure can be more reliably monitored compared to change in electric signals that appear on skin. In this study, the suggested sensing method's principles and applications are explained. Motion sensor & EMG signal are used to compare the performance of the suggested approach like repeatability, linearity, etc.

Adam B. Noel [11] gave a detailed assessment of SHM employing WSNs in 2017, explaining the algorithms utilized in damage detection and localization, outlining network design problems, and defining future research goals. Many network design issues are examined and explored, including scalability, time sync, sensor location, and data processing. Additionally, this survey includes a summary of WSN testbeds and real-world installations for SH.

For remote health monitoring applications, Ravi Kiran [13] developed an innovative smart transmission approach with a smooth handoff mechanism in 2015 that uses several on-chip radios. An adaptive rule engine is used to categories the acquired multi-parametric data from the patient and communicate, it only when it is required. As part of the on-chip seamless handoff mechanism described, the intelligent regulation of the various on-chip radios helps to provide ubiquitous connection while conserving a significant amount of battery power. On average, the proposed on-chip seamless handoff mechanism and adaptive rule engine-based smart transmission mechanism save 50% of energy and reduce the transmitter's duty cycle by 51% when compared to continuous transmission, according to performance study done with 20 users. The suggested seamless handoff controller and adaptive rule engine only need 2082 CMOS transistors for real-time implementation, according to an examination of their hardware complexity.

For a BSN-based contemporary healthcare system, Gopi [14] outlined the most important security criteria in 2016. As a result, we've developed BSN-Care, an IoT-based healthcare system that makes use of BSN to meet these needs. One of the key IoT breakthroughs in healthcare is the body sensor network (BSN), which uses a collection of small, light, and powered wireless sensor nodes to keep tabs on a patient. This new technology poses a threat to patient privacy since it was used in medical settings without adequate security measures in place.

When blood is pumped from the heart to the head, minor oscillatory head movements owing to Newtonian mechanics may be identified for periodicity analysis in 2017. The authors in [15] proposed to collect depth footage of a human subject using Kinect 2.0 to estimate his/her heart rate and rhythm. First, we use graph-signal smoothness prior to regularization to recover an acquired depth video using a combined bit-depth improvement / denoising process. Secondly, we use the depth video to monitor an area of the head that is automatically recognized in order to derive 3D vectors of motion. In order to guarantee that the motion information in the two modules is constant, the detected vectors are given back to the depth restoration module in a loop. This improves the efficiency of both restoration and motion tracking. There are three stages in this process: first are trend elimination, then band-pass filtering and wavelet-based motion denoising for 1D signal analysis. Finally, Welch power spectrum analysis is used to measure the heart rate and peak detection is used to calculate the heart rhythm. A portable oximeter's heart rate and rhythm estimates were shown to be inaccurate in comparison to our algorithm's heart rate and rhythm estimates.

In 2022, BharathWaj [16] presented a research & analysis of 3-D human body location & tracking utilizing IR-UWB technology, which is small and cost-effective to wear on the body. Analysis of body-centric wireless channel characterization using criteria like path loss magnitude, and Kurtosis to distinguish between LOS and non-LOS circumstances has been carried out in detail. Studies on the pulse preservation properties of the UWB antenna while worn by a human subject include fidelity calculations for various activities and placements of the antenna. Localization accuracy of 90% is reported in this article, which is equivalent to findings produced by conventional optical motion capture system, employing simple and cost-effective procedures.

3. Problem Statement

There are many researches in field of sentiment analysis but there is need to introduce approach where real life issues should be considered. Limited researches have considered mental stress detection and analysis. Moreover there is need to propose mechanism that should be capable to provide high accuracy along with performance.

4. Objectives

The objectives of research are discussed below:

1. To consider the role of deep learning in considering the sentiments to detect and analysis mental stress in online learning.
2. Considering issues in sentiment detection and resolving issues that are arosen during sentiment detection and analysis.
3. To propose novel approach in order to train machine by considering feedback from people after survey.
4. To evaluate the performance and accuracy of prediction and analysis model.

5. Research Methodology

Training in scientific instrument and technique selection, as well as materials for putting research into practice, is provided through research methodology. The following sections provide an overview of many types of research methodologies: To collect, select, analyze, and evaluate data on a certain topic, researchers use techniques. The methodology section of a research report may help readers determine the study's overall validity and dependability. The current research methodology is shown in Figure 7.

- *Problem identification:* We are considering the issues of mental stress and factor identification.
- *Literature survey:* We are viewing the Online Learning, Sentimental Analysis and IoT Devices used in previous researches.
- *Proposing novel approach:* We are implementing the detection and analysis of mental stress and also considering the sentiment of user.
- *Evaluation:* Performance and accuracy of our proposed work are comparing with the previous work.

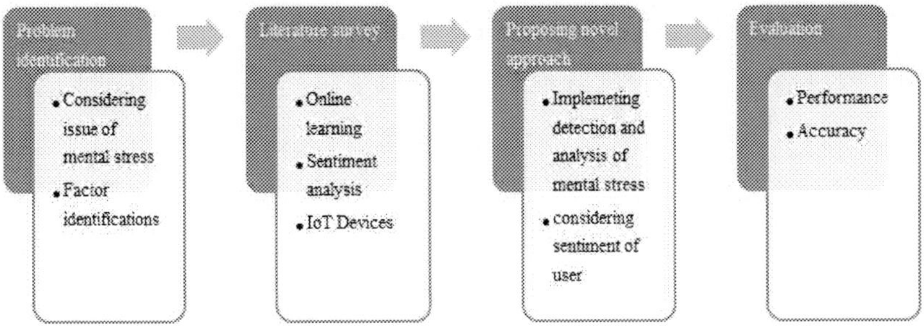

Figure 7. Research methodology.

6. Proposed Work

Proposed work (as described in Figure 8) is based on sentiment analysis mechanism [36] that is considering detection and analysis of mental stress. Such analysis and detection take place in online learning environment. IoT devices are used to collect the dataset of real-word [37, 38] in order to perform detection and analysis of mental stress.

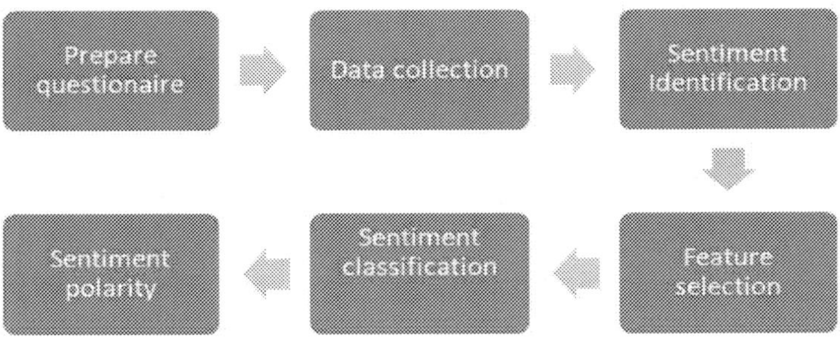

Figure 8. Proposed model.

During survey feedback from users are considered and sentiment identification process is implemented. Then the sentiment classification is performed after feature selection. Finally sentiment polarity takes place. Data set of 10000 records is considered for sentiment classification.

7. Result and Discussion

Records of 10000 records have been considered where 7000 records are taken for training and 3000 out of them have been taken for testing. Matrix that is presenting total negative, neutral and positive sentiments is classified in Table 2.

Table 2. Overall classifications of data for testing

	Negative	Neutral	Positive
Negative	1300	0	0
Neutral	0	800	0
Positive	0	0	900

7.1. Confusion Matrix Obtained in Case of Conventional Deep Learning

Table 3. Confusion matrix obtained in case of conventional deep learning

	Negative	Neutral	Positive
Negative	1102	154	44
Neutral	168	556	76
Positive	80	225	595

Result
TP: 2253
Overall Accuracy: 75.1%.

Table 4. Metrics obtained in case of conventional deep learning

Class	n (truth)	n (classified)	Accuracy	Precision	Recall	F1 Score
1	1350	1300	85.13%	0.85	0.82	0.83
2	935	800	79.23%	0.69	0.59	0.64
3	715	900	85.83%	0.66	0.83	0.74

7.2. Confusion Matrix in Case of Proposed Work

Table 5. Confusion matrix in case of proposed work

	Negative	Neutral	Positive
Negative	1196	90	14
Neutral	124	624	52
Positive	51	124	725

Result
TP: 2545
Overall Accuracy: 84.83%.

Table 6. Metrics obtained in case of proposed work

Class	n (truth)	n (classified)	Accuracy	Precision	Recall	F1 Score
1	1371	1300	90.7%	0.92	0.87	0.90
2	838	800	87%	0.78	0.74	0.76
3	791	900	91.97%	0.81	0.92	0.86

7.3. Comparison Analysis

Comparison of traditional and proposed F1 Score, accuracy, precision and recall in case of previous work and proposed work is shown in Table 7.

7.3.1. Accuracy

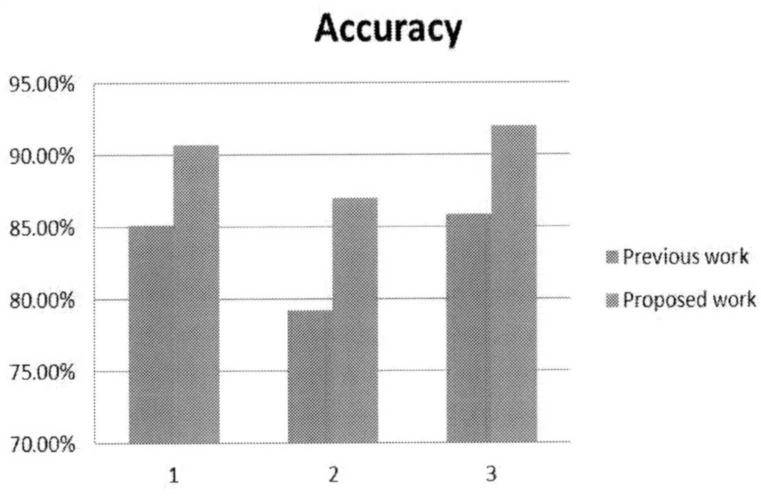

Figure 9. Comparison of accuracy.

Table 7. Comparative analysis of accuracy

Previous work	Proposed work
85.13%	90.7%
79.23%	87%
85.83%	91.97%

7.3.2. Precision

Table 8. Comparative analysis of precision

Previous work	Proposed work
0.85	0.92
0.69	0.78
0.66	0.81

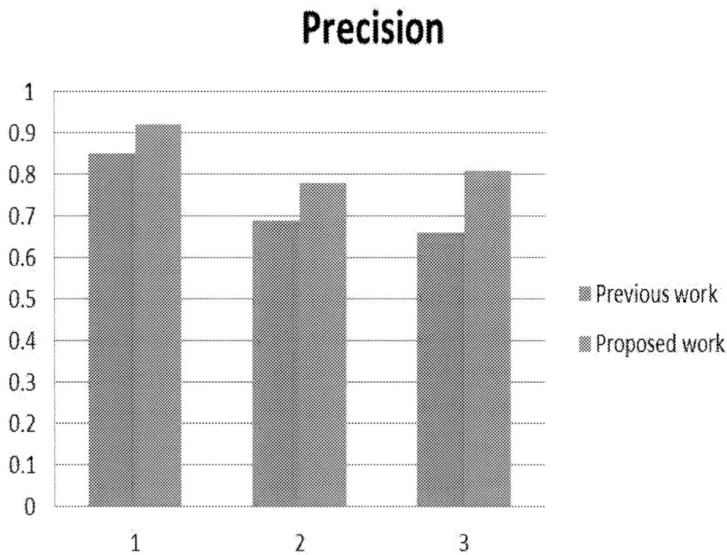

Figure 10. Comparison of precision.

7.3.3. Recall Value

Table 9. Comparative analysis of recall value

Previous work	**Proposed work**
0.87	0.87
0.74	0.74
0.92	0.92

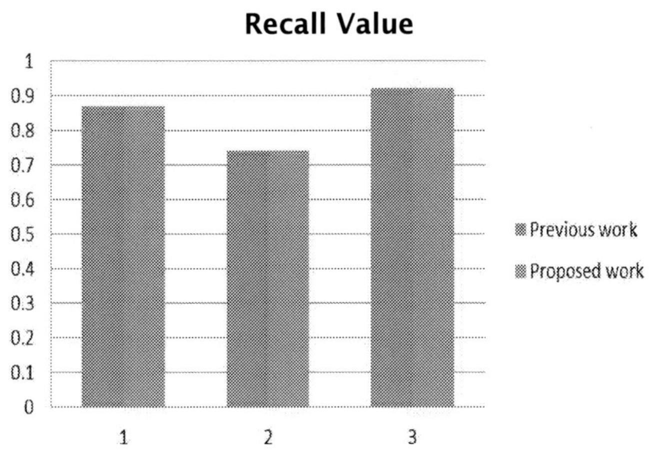

Figure 11. Comparison of recall value.

7.3.4. F1-Score

Table 10. Comparative analysis of F1-Score

Previous work	Proposed work
0.83	0.90
0.64	0.76
0.74	0.86

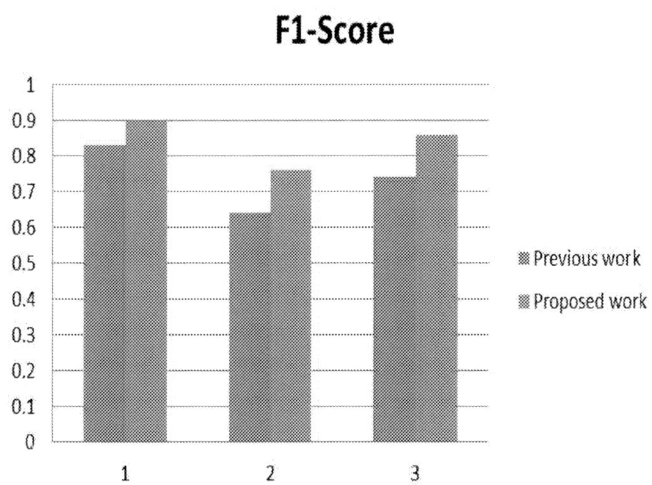

Figure 12. Comparisons of F1-Score.

Conclusion

Mentalstress is wreaking havoc on both our mental and physical well-being. A person's level of stress might be difficult to discern from their behavior and facial expressions. It's possible that the individual may seem to be in better condition, but their body is still under a lot of stress. Constant stress has the potential to cause long-term damage to their mental well-being, which might lead to a variety of chronic health problems. Regular stress monitoring, on the other hand, is critical for spotting health problems that might develop into long-term illnesses down the road. Wireless networks based on IOT (Internet of Things) enable a broad variety of chances to periodically monitor stress levels and convey information to the concerned for urgent action. Various sensors, such as heart rate, blood pressure (BP), body temperature, and CO_2 gas concentration, are used in the development of a model to identify stress. It is also computed based on these sensors' readings that the levels of stress are sent through IOT.

Future Scope

In future, studies couldbe conducted in the area of sentiment analysis, but a strategy that takes into account real-world challenges is needed. Only a few studies have looked at the detection and analysis of mental stress. In addition, a technique that is both accurate and efficient is needed to be proposed.

References

[1] Gupta, A., Srivastava, A., and Anand, R. (2019). Cost-effective Smart Home Automation using Internet of Things. *Journal of Communication Engineering and Systems,* 9(2), 1-6.

[2] Anand, R., Mishra, A., and Gupta, R. (2020). IoT based Smart Energy Meter using NodeMCU. *Test Engineering and Management,* 82, 4297-4301.

[3] Bakshi, G., Shukla, R., Yadav, V., Dahiya, A., Anand, R., Sindhwani, N., and Singh, H. (2021). An Optimized Approach for Feature Extraction in Multi-Relational Statistical Learning.*Journal of Scientific and Industrial Research (JSIR),* 80(6), 537-542.

[4] Kohli, L., Saurabh, M., Bhatia, I., Sindhwani, N., and Vijh, M. (2021). Design and Development of Modular and Multifunctional UAV with Amphibious Landing,

Processing and Surround Sense Module.Unmanned Aerial Vehicles for Internet of Things (IoT) Concepts, Techniques, and Applications, 207-230.

[5] Gupta, A., Anand, R., Pandey, D., Sindhwani, N., Wairya, S., Pandey, B. K., and Sharma, M. (2021). Prediction of Breast Cancer Using Extremely Randomized Clustering Forests (ERCF) Technique: Prediction of Breast Cancer.*International Journal of Distributed Systems and Technologies (IJDST)*,12(4), 1-15.

[6] Srivastava, A., Gupta, A., and Anand, R. (2021). Optimized smart system for transportation using RFID technology.*Mathematics in Engineering, Science and Aerospace (MESA),* 12(4).

[7] Kashif, M., Javed, M. K., and Pandey, D. (2020). A surge in cyber-crime during COVID-19. *Indonesian Journal of Social and Environmental Issues (IJSEI),*1(2), 48-52.

[8] HealthGare: "A Real-Time Wearable System for Monitoring and Analyzing Physiological Signal" Nuria Oliver 1 Fernando 2008.

[9] Arman, University of Turku, "HICH: Hierarchical Fog-Assisted Computing Architecture for Healthcare IOT" *ACM Transactionson Intelligent Systems and Technology,* Vol.1, No.4, September 2017.

[10] Iman, Auburn University, "Tensor Beat: Tensor Decomposition for Monitoring Multi-person Breathing Beats With Commodity WIFI" *ACM Transactionson Intelligent Systems and Technology,* Vol.5, No.2, September 2017.

[11] Adam B. Noel, "Structural Health Monitoring Using Wireless Sensor Network," *IEEE Transactions On Biomedical Engineering,* Vol.17, No.3, February 2017.

[12] Junk, StudentMember, "A Wearable Gesture Recognition Device for Detecting Muscular Activities Based on Air-Pressuresensors" *IEEE Transactions on Industrial Engineering,* Vol.5, No.8, April 2016.

[13] RaviKiran, "System Architecture For Low Power Ubiquitously Connected Remote Health Monitoring Applications With Smart Transmission Mechanism" *IEEE Transactions On Sensors,* Vol.15, No.8, August 2015.

[14] Gopi, "BSN–Care: A Secure IOT Based Modern Health Care System Using Body Sensor Network" *IEEE Transactions On Sensors* Vol.6, No.5, March 2016.

[15] Velnath Ravanan, R. Subasri, M. G. Vimal Kumar, K. T. Dhivya, P. S. Kumar and K. Roobini, "Next Generation Smart Garbage Level Indication and Monitoring System using IoT," *2021 Smart Technologies, Communication and Robotics (STCR),* 2021, pp. 1-4, doi: 10.1109/STCR51658.2021.9588961.

[16] BharathWaj, "Impulse Radio Ultra-Wide band Communications For Localization and Tracking of Human Body and Limbs Movement for Healthcare Applications" *IEEE Transactions On Antennas and Propagation,* Vol.65, No.12, December 2017.

[17] Hamjisa, "Design and implementation of an Accurate, Portable and Time – Efficient Impatient – Based Transceiver for Structural Health Monitoring" *IEEE Transactionson Mechatronics,* Vol.2, No.6, December 2017.

[18] Woo Kim, "Heart Rate Detection During Sleep Using a Flexible RF Resonator and Injection – Locked PLL Sensor" *IEEE Transactions On Biomedical Engineering,* Vol.62, No.11, November 2015.

[19] Lin, "A Non-Contact and Cost- Effective Sleep Monitoring System" *IEEE TransactionsonBiomedicalCircuits and Systems,* Vol.4, No.1, February 2017.

[20] Omran, M. A., Saad, W. K., Hamza, B. J., and Fahe, A. (2021). A survey of various intelligent home applications using IoT and intelligent controllers. *Indonesian Journal of Electrical Engineering and Computer Science,* 23(1), 490–499. https://doi.org/10.11591/ijeecs.v23.i1.pp490-499.

[21] Khanna, A., and Kaur, S. (2020). Internet of Things (IoT), Applications and Challenges: A Comprehensive Review. In *Wireless Personal Communications* (Vol. 114, Issue 2). Springer US. https://doi.org/10.1007/s11277-020-07446-4.

[22] Yang, T., Zhang, G. H., Liu, L., and Zhang, Y. Q. (2020). A survey on authentication protocols for internet of things. *Journal of Cryptologic Research,* 7(1), 87–101. https://doi.org/10.13868/j.cnki.jcr.000352.

[23] Fadhil, S. A. (2021). Internet of Things security threats and key technologies. Journal of Discrete Mathematical Sciences and Cryptography, 2021. https://doi.org/10.1080/09720529.2021.1957189.

[24] Wang, D., Liu, H., Ma, X., Wang, J., Peng, Y., and Wu, Y. (2017). Energy harvesting for internet of things with heterogeneous users. *Wireless Communications and Mobile Computing,* 2017. https://doi.org/10.1155/2017/1858532.

[25] Abdelgawad, A., and Yelamarthi, K. (2017). Internet of Things (IoT) platform for structure health monitoring. *Wireless Communications and Mobile Computing,* 2017. https://doi.org/10.1155/2017/6560797.

[26] Ramirez, A. R. G., González-Carrasco, I., Jasper, G. H., Lopez, A. L., Lopez-Cuadrado, J. L., and García-Crespo, A. (2017). Towards Human Smart Cities: Internet of Things for sensory impaired in dividuals. *Computing,* 99(1), 107–126. https://doi.org/10.1007/s00607-016-0529-2.

[27] Pandey, D., Ogunmola, G. A., Enbeyle, W., Abdullahi, M., Pandey, B. K., and Pramanik, S. (2021). COVID-19: A framework for effective delivering of online classes during lockdown.*Human Arenas,* 1-15.

[28] Ayenew, B., Pandey, D., Yitayew, M., Etana, D., Binay Kumar, P., and Verma, N. (2020). Risk for surge maternal mortality and morbidity during the ongoing corona virus pandemic.*Med Life Clin,*2(1), 1012.

[29] Parthiban, K., Pandey, D., and Pandey, B. K. (2021). Impact of SARS-CoV-2 in online education, predicting and contrasting mental stress of young students: a machine learning approach.*Augmented Human Research,*6(1), 1-7.

[30] Babu, S. Z. D., Pandey, D., and Sheik, I. (2020). Acts Of COVID19.*International Journal of Advanced Engineering, 3,* 2457-0397.

[31] https://www.electronicscomp.com/image/cache/catalog/original-arduino-uno-rev3-400x400.jpg.

[32] https://techblog.livongo.com/wp-content/uploads/2018/05/wrist_bpm.jpg.

[33] GSM/GPRS Module with RS232 Interface – IOT, EMBEDDED TRAINING, CIRCUIT DESIGN (turjasuzworld.in).

[34] https://www.aliexpress.com/item/32909924229.html.

[35] Buy ESP8266 NodeMCU CP2102 Board at Lowest Price In India | Robu.in.

[36] Jain, S., Sindhwani, N., Anand, R., and Kannan, R. (2022). COVID Detection Using Chest X-Ray and Transfer Learning. In*International Conference on Intelligent Systems Design and Applications* (pp. 933-943). Springer, Cham.

[37] Meivel, S., Sindhwani, N., Anand, R., Pandey, D., Alnuaim, A. A., Altheneyan, A. S., ... and Lelisho, M. E. (2022). Mask Detection and Social Distance Identification Using Internet of Things and Faster R-CNN Algorithm.*Computational Intelligence and Neuroscience,* 2022.

[38] Jain, S., Kumar, M., Sindhwani, N., and Singh, P. (2021, September). SARS-Cov-2 detection using Deep Learning Techniques on the basis of Clinical Reports. In*2021 9th International Conference on Reliability, Infocom Technologies and Optimization (Trends and Future Directions) (ICRITO) (pp. 1-5). IEEE.*

Chapter 9

COVID-19: The Enemy Threatening Mental Health among Medical Warriors

Abrar Bakry Malik[1,*] and Mohamed Eltayieb Elawad[2]

[1]Faculty of Medicine and Health Science, Omdurman Islamic University, Khartoum, Sudan

[2]Faculty of Medicine, National Ribat University, Khartoum, Sudan

Abstract

Corona virus disease – 19 (COVID-19) is one of the most prominent pandemics of the 21st century. Since its appearance in China, it had a lot of bad consequences in various aspects. In this chapter, we will talk about the loads imposed by the nature of work in the health facilities during the pandemic, and the significant psychological diseases caused by those burdens, citing some of the cross-sectional studies that have been done around the world. Also, we touched on the groups most vulnerable to those disorders and we have outlined the most important strategies for dealing with the crisis.

Understanding the stressors and psychological effects of COVID-19 is one of the battle rounds that we can easily gain with our rational coping with mental problems.

Keywords: COVID-19, mental health, psychological disease, depression, anxiety, stressors, and healthcare workers

[*] Corresponding Author's Email: bakry.abrar287@gmail.com.

In: An Interdisciplinary Approach in the Post-COVID-19 Pandemic Era

Editors: D. Pandey, B. Kumar Pandey, N. Sindhwani et al.

ISBN: 979-8-88697-052-4

1. Introduction

Mental health is to realize that one can cope with the pressures of life, can function efficiently, and make a positive contribution to society. It is clear from the above that world health organization (WHO) has relied on the production aspect to formulate its definition. But other aspects have not been addressed that represent the moral aspect of mental health which are psychological factors like complacency, good relations with others and the ability to manage day-to-day activities effectively, and emotions such as the desire for life, happiness, satisfaction, and passion. So if we want to define mental health comprehensively, we need to integrate all the aspects mentioned [1-3].

The global burden of mental illness has reached proportions that cannot be underestimated. In 2016, the negative impact it reflects on the quality of life has been estimated using years of life with disability (YLDs) and disability-adjusted life years (DALYs) at 32.4%, and 13%, respectively [4].

COVID-19 is an infectious disease caused by the severe acute respiratory syndrome-corona virus-2 (SARS-CoV-2) virus [5, 6, 7, 8]. The outbreak appeared in late December 2019 and was declared a pandemic on 12 March 2020. It often causes respiratory symptoms like cough, shortness of breath, and sore throat. But the recent researches have shown that it can cause clinical manifestations of biological systems other than pulmonary one. Features range in intensity from mild to moderate. However, it can be severe or critical, thus patients need intensive care units if the manifestations of acute respiratory distress syndrome occur [9].

With each wave of COVID-19, the morbidity and mortality rates are rising worldwide, as provided for in March 2020 statistics, the cumulative infected cases are 116,166,652 and the cumulative of deaths is 2,582,528 [10].

COVID-19 pandemic has had a major impact on various aspects of life such as health, economy, and agriculture [11, 12, 13]. Many countries have tried to reduce the spread of infection by various means. Those have tended to lockdown completely or partially or apply both interchangeably. Also, they forced people to wear protective masks in public utilities, commit to social distance, and wash their hands for at least 60 seconds in a particular technique repeatedly.

The health sector is the first to be affected by the pandemic. Health systems have fallen, medical resources have been reduced, and the numbers of infections and death have worsened terribly. That all had a serious impact on the psychological integrity of the communities generally.

To illustrate the psychological impact of COVID-19 pandemic, we used this systematic review and meta-analysis which indicates that the prevalence of depression during the pandemic is 25% [14].

To be clearer, researchers compare the pre and post-pandemic rates of the same disorder in a highly infected country and another one with low rates of infection. It was three times higher during COVID-19 in the United States of America [15].

In Cyprus, the prevalence of depression was only 11.5% before the pandemic, and now it is about half of the population (48%) [16, 17].

Healthcare workers are a group subject to psychiatric illness, 48.6% of them are found to suffer from common psychiatric disorders (CPD). That is expected, due to the stressful nature of their work [18].

In this chapter, we will discuss the main stressors faced by medical staff during the COVID-19 pandemic and the resulting psychological disorders. Also, we will review the management methods that reduced the psychological impact.

2. Stressors vs Health Workers

The medical personnel faced many challenges that posed a ghost that threatened their safety. Between the maintenance of personal safety and the provision of optimal care for patients, the pressure is increasing and is exacerbated by the scarcity of personal protective equipment (PPEs) in many countries. Italy is one of the countries that has formed the scene of this tragedy with peak rates of infection and death. The United States of America continues to advocate its need for more ventilators and protective masks [19, 20].

China is the largest manufacturer and exporter of PPEs, at the same time it is the source of SARS-CoV-2 and one of the countries with high rates of morbidity and mortality. Therefore, the equipment will not be deployed outside unless self-sustainment is achieved, whose ceiling has also been increased by higher internal demand [21].

With a lack of gowns, N-95 masks, gloves, eye protectors, and face shields, some countries are unable to protect frontline hospital workers, even after internal production has increased to its maximum capacity.

Moreover, the available equipment is not preferred for reuse, as a study has shown that doctors who use PPEs more than once have a double chance of getting infected compared to those who only use it once [22].

All of these have forced some authorities to import low-quality equipment from unregistered entities to fill the deficit. Because the quality is low, the deficit has not been filled and medical staff prevention has not been achieved completely [22].

We need to adopt some strategies that can alleviate this obsession with medical personnel. First, all countries, especially those with few local factories, should provide facilities and support to businessmen to increase national medical equipment manufacturers. Secondly, to ensure the fair distribution of resources, the ministries of health should give hospitals with high infection intensity large quotas of medical and preventive supplies. Thirdly, healthcare workers must follow the center for disease control (CDC) and WHO guidelines of PPEs consumption and avoid non-rational use [23].

Fourthly, the various media must encourage and inform local and international communities about the importance of donating to the health sectors and exchanging reinforcements, so that we can safely overcome the pandemic.

The increased workload is one of the factors that strain mental energy in various professions, particularly medical practice as many studies have shown that workers in this field bear enormous pressure compared to others. These studies have been done in normal working conditions, so it is expected that workload will increase twice as much in an event as the COVID-19 pandemic [24, 25].

Many risk factors have been identified in the negative impact on mental health, primarily long daily working hours that may be 24/24 hours, anxiety in the female staff, and associated factors such as body mass index (BMI) with anxiety and insomnia [24].

In my view, the flood of infected and critically ill patients to the extent that neither hospital's equipment nor the medical personnel can absorb it, the collapse of health systems, and the weakness of resources are also factors that have somehow weighed the burden.

To deal well with mounting work pressures, we should appreciate the reality imposed on medical crews. They are facing two surprising ends. The first is the doubling of their tasks, and the other is the shortage of medical resources. Also, they have the burden of being the patient's family and showing the human side to him throughout his treatment journey at the health facility. We do not lose sight of the household and community responsibilities that he has to accomplish after work, which can often be at the expense of his rest time.

Training cadres in crisis and time management may be one effective solution, which could be a proposal for a study exploring the impact of this training on healthcare workers. Such training should target affected countries in general, and high-income ones in particular, whose crews have been shown to experience high burnout rates compared to low-income ones [25].

Normally, medical professionals are at risk of infection. But, it gets terrifying when every patient you deal with is a source or transmitter of infection. It is a very stressful condition. Fear of infection is a barrier between the health care provider and the patient and may lead him to leave the health facility at all. This fear is an intermediate point in the sequence of problems faced by warriors of coronary viruses' invasion, which is an extension of the scarcity of PPEs that we mentioned earlier, and a start to the dilemma of personnel working in the front lines shortage that will end up collapsing the health system if this chain is not broken. In a study conducted in Egypt, doctors were found to suffer from extreme fear of COVID-19, and this fear was directly related to the depletion of their mental health, exhaustion, and low quality of their lives. It's not just about this, it's about affecting COVID-19 patients and others, as a study has shown that negligence of suspected patients by health workers led them to suicide [26, 27].

Fear of COVID-19 among health workers can be caused by the severe complications and consequences shown by many cases. The clinical presentation of SARS-CoV-2 is not necessarily respiratory-centered but can have different systemic features like digestive, renal, or cardiovascular manifestations. So misdiagnosing of the disease can happen and heighten doctors' fears. On the other hand, an individual can be a silent carrier that transmits the infection to his or her surroundings. Family is an important part of this surrounding, and the transmission of infection to its members is the major fear of frontline workers [28, [29].

The stigma of COVID-19 infection also has a role to play in raising workers' fears and avoiding workplaces. Moreover, females, clinical experience for five years or less, and house officers were found to be overly fearful compared to others in the same setting [30, 31].

To reduce the fear of infection, medical administration should assure health workers that they will not be in the frontline alone and that they will provide them with all the support necessary to overcome the crisis. That support should be preceded by a psychometric assessment of fear. According to its result, the support is directed to the most vulnerable individuals [32].

Also, administrations should keep health workers informed of the medical practice modifications during COVID-19 pandemic continuously, and make them aware that these guidelines have been decided to keep them safe.

The stigma of infection is a misconception in societies, and the fear of health workers is one of its most serious effects. So if we want to get rid of the impact, we must correct the effector by seeking to make society aware of the misconception and the resulting negative manifestations in a scientific manner appropriate to different environments and educational backgrounds.

Social distancing is one of the protective measures of COVID-19 combating. WHO has set one to one and a half meters for social distancing [33].

It is working on the principle of breaking the chain of infection, thereby reducing its spread in society. Although it seems to be of great use, it does have negative effects on mental health. In this paper, social distancing is related to depression, drug addiction, and post-traumatic stress disorder (PTSD) [34].

The human being tends to communicate socially. That strengthens the sense of interdependence and makes him feel that he is a part of the surrounding society that affects and is influenced by it. Thereby promoting feelings of acceptance and safety. When social distancing is applied, those feelings may fade away, and the distancing will shift from preventive mean to stressful factor.

In the same context of social distancing, social isolation is one of the stressors that plays a central role in creating psychological problems. World Health Organization has determined an isolation period of 14 days, which is enough time to eliminate the virus from human serum, while at the same time it is enough to break down the infected person and make him feel untouchable [35].

It's a matter that casts a dark shadow on the mental health of the patient. In this study, social isolation is linked to stress, anxiety, fear, and low mood [36].

Being sick and exhausted without direct moral support from relatives makes an isolated patient feels excluded and lonely, which may weaken his resistance and response to treatment.

Despite the psychological effects, we have to balance the preventive benefit of social distancing and isolation, and community stability. It is difficult to abandon these protective measures at this time, especially there is still no specific therapeutic option for eliminating coronary viruses. In addition, they are easy to apply, and without cost. So, excluding them means resorting to other alternatives that may be of high expense that the economies of the affected countries cannot afford. At the moment, we must maintain social distancing and isolation as essential defense methods, coupled with the provision of necessary education and psychological counseling.

3. Mental Disorders: Cross-Sectional Studies Conducted Worldwide

It is a matter of the promise that during those events the world heeds the psychological impact of the pandemic. Several cross-sectional studies have been conducted targeting several aspects of mental health. Depression and anxiety are major psychological problems facing defensive line workers. In France and the United Kingdom, they suffer severely compared to others in other fields [37].

Also, India performed a survey for 152 doctors using the depression, anxiety, and stress scale - 21, and concluded that 34.9% are depressed, and 39.5% are anxious. In the same context, it has been concluded that those performing clinical duties face a higher risk of depression and anxiety (20% and 23.9%) than others (14.2% and 15.5%), respectively [38, 39].

Hunan Province, China was a pioneer of psychological interest, singling out a study of neurology doctors and nurses and found that depression and anxiety in nurses are higher than in doctors. In the same country, anxiety and depression are high in most professions (30.3% and 17.9%) in general, other medical personnel than doctors and nurses, and students in particular [40].

Within 65 Pakistani cities, depression/anxiety was found to be 43% among doctors. For workers in intensive care units it was concluded that they were depressed and severely anxious by 6%, and 11%, respectively [41, 42].

This is not very different when talking about stress and post-traumatic stress disorder (PTSD), as 40% of intensive care cadres perceive severe post-traumatic stress symptoms [42].

It was found that the average doctor suffering stress is 20.6 according to the perceived stress scale -10 (PSS-10). In Italy, COVID-19 health workers showed symptoms that often refer to PTSD. Also, 16.7% of the medical staff had PTSD in Greece [43, 44, 45].

The mental impact has not only been the preserve of countries that have recorded high morbidity and mortality rates, but even low-infected ones such as Cyprus also have not been immune [46].

The tragic end of COVID-19 is the occurrence of acute respiratory distress syndrome which often leads to death, where complicated grief happens. The estimated prevalence is 5% to 52% among the families of the dead patients in intensive care units [47].

Relatives of ICU residents are found that 46% of them suffered from complicated grief and PTSD after 6 months [48].

In Italy, 933 individuals from various medical personnel were surveyed. Somatization disorder was found to be most prevalent at 71% [49].

Healthcare workers are our front lines of defense on which everyone's hopes are placed. Because psychiatric disorder may be interpreted as a type of vulnerability and considered as a stigma rather than a disease, different stress factors are translated into physical symptoms as being more expressive of suffering.

In general, restrictions imposed on physical integrity have had negative effects on mental health. Suicide is the worst of them. It is estimated to have risen from 1% to 145% during COVID-19 [50].

Directly, it could be the end of the sad sequence of events of the epidemic or those painful memories have tightened their grip on high-risk individuals. In the medical field, the chance of thinking about self-harm is twice as high as those in the general population. In addition, suicide rates among doctors are the highest compared to those in other fields [51].

Once again, that can be attributed to increased work pressures and the extremely complex nature of the pandemic. Furthermore, medical practitioners may be ashamed and reluctant to seek psychiatric counseling early.

The alarming suicide rates call for preventive interventions by performing periodic surveys among healthcare workers to assess the existence, and severity of the psychiatric diseases and risk of suicidal ideation. Also, we have to deal with high-risk individuals with a high level of responsibility for their work environment, and their living conditions outside the walls of therapeutic units. These are essential actions to prevent the risk of suicide.

However, other cross-sectional studies have highlighted other mental illnesses that medical staff facing COVID-19 suffer from, such as psychological trauma and panic attacks [52, 53].

In a way, the stressors are affecting the mental integrity of health care providers working in COVID-19 wards. Some studies have been conducted to understand risk factors or causes that can precipitate psychiatric problems higher. Some of these findings were obtained in India, which agreed with what was mentioned in the stressors paragraph, such as lack of PPEs, and long hours of work. Also, it was reported that a sacrificing attitude for patients might be one of the reasons [38].

Chinese researchers have concluded that junior cadres, and doctors who are under the age of 40, are at a higher risk of depression and anxiety overall. In addition, lack of PPEs is particularly responsible for anxiety, and young age is specifically related to depression [40].

In contrast, the findings of a study done in 8 European countries suggested that working conditions and coping with infected patients had nothing to do with mental integrity [37].

Medical personnel is responsible for countering COVID-19. This task requires them to show cohesion and rigidity to complete it perfectly. This means that various pressures have been suppressed, as well as the need to preserve the image of a steadfast warrior whose hopes of the patients and their families are related to. It all comes together to be translated into the aforementioned psychological disorders.

We can attribute the high rates of complicated grief to the sudden emergence of the outbreak and the sudden loss of a huge number of lives. Moreover, the strict applied precautionary measures, primarily social distancing and social isolation, may cause relatives to blame themselves and feel remiss as they have not even been able to say goodbye or prepare a funeral following their religion and tradition [47].

The high prevalence rate of somatization is a risk predictor that management methods are needed to reduce. So, researchers have done a lot of studies to assess the efficacy of some of those methods. Reducing stressors, health training and education, and coping with difficult conditions decreases the incidence of somatization disorder by 5% [54].

Also, there are many daily activities like practicing hobbies, playing visual and mental games, doing domestic work, and exercising play a prominent role in reducing the chances of somatization disorder [54].

Some health workers thought suicide was a radical solution to end their suffering. That's what emergency doctors in New York, one medical student, paramedics, and some nurses already have. Most of those who did this were found female and between the age of 22 and 60. Moreover, they are Indian [55, 56].

4. Management

COVID-19 pandemic and its psychological effects require us to implement integrated strategies to preserve the mental health of medical workers. These strategies should not be the responsibility of a particular authority or individual, but a shared responsibility of three essential elements: the individual, peers & colleagues, health facilities, and governments.

4.1. Individual

An individual must act to protect himself from the psychological threats posed by the pandemic, especially if he or she is one of those classified as a high-risk for psychological disorders precipitation. Therefore, he should give himself sufficient time to rest physically and mentally, and he should reduce the altruistic behavior that can destroy his mental well-being and lead him to burnout [57, 58, 59]. Also, he must not lose sight of good food, drink, exercise, and useful mind games, focusing on positive thinking and continuing to look at the bright side of the situation [60, 61].

Additionally, he has to communicate regularly with his family, friends, and inspiring people in his life. He has to talk about fascinating topics with them and avoid that about work as much as possible. Moreover, he must be proud of the pivotal role he plays and is seen with passion and satisfaction, and do not limit to providing medical services at his facility. He should educate his community about COVID-19 drawing on his experience.

4.2. Peers and Colleagues

Cooperation between colleagues drives the work forward and promotes the concept of mutual psychological protection among the medical teams. The most important thing for colleagues to play a major role in preventing psychiatric illness is good communication [62].

The latter breaks the barriers, relieves the pressure of the pandemic, facilitates the disclosure of suffering, and reduces the sense of loneliness.

Also, peers know the strengths and weaknesses of each other. Therefore, they can provide moral and sensible support in a way that enhances their strengths and improves their weak points. In the same context, that knowledge helps them to distribute tasks fairly among themselves so that each member is placed in his proper place and performs duties appropriately.

In the line of duty, a member of the medical staff goes through a situation that may weaken his resolve or make him feel frustrated or angry, and that's where the members need each other. They must understand humanity and appreciate the hard circumstances of the crisis. Also, they must absorb anger and negative emotions from each other and throw them out, breathe team spirit, and encourage each other that they are all capable of overcoming the crisis.

Moreover, colleagues are the closest surrounding of a health care provider. That's why they're noticing any changes of concern to their colleagues, like excessive intolerance, irritability, nightmares, or any other physical symptoms. In this case, their role is advice, reassurance, and turn him into a psychologist or a physician suitable for his condition, rather than observation.

4.3. Health Facilities and Governments

Medical administrations and governments have the highest authority in the prevention of the psychological effects of COVID-19. They must be aware and address the vast majority of stressors and psychological disorders. Also, they should provide the best services and programs to medical staff in general, and high-risk individuals in particular.

Hospitals should provide their staff with a configured environment in terms of healthy nutrition and drinks, resting and sleeping spaces, entertainment, internet, and means of online communication [60].

Also, major resources should be available, such as the presence of expert doctors and specialists in the field of immunology and infection control, provision of more intensive care units and emergency rooms, and establishment of more psychiatric clinics to provide health workers with counseling and guidance to cope with workload properly. The safety of medical staff must be a priority, and that is translated into an abundance of PPEs, application of precautionary measures, and persistence of strict disinfection and sterilization.

Working hours organization by adoption of shifts system and duties distribution fairly and comfortably are of the utmost importance [63].

Conclusion

The role of administrations and governments should not be limited to health facilities but should be concerned with workers' families and home safety. Also, they have to provide well-equipped transport vehicles that bring staff from hospitals to their homes to ensure their safety and that of the community.

Furthermore, education programs and training sessions are supposed to be conducted for health workers on coping strategies for stress and mental diseases.

All that has been mentioned can be crowned by meaningful dialogue between administrations and staff, and acceptance of opinions and suggestions that improve working conditions and the quality of medical services provided.

References

[1] Galderisi, S., Heinz, A., Kastrup, M., Beezhold, J., & Sartorius, N. (2015). Toward a new definition of mental health. *World Psychiatry*, *14*(2), 231.

[2] World Health Organization. (2004). *Promoting mental health: Concepts, emerging evidence, practice: Summary report.*

[3] Keyes, C. L. (2006). Mental health in adolescence: Is America's youth flourishing?. *American journal of orthopsychiatry*, *76*(3), 395-402.

[4] Vigo, D., Thornicroft, G., & Atun, R. (2016). Estimating the true global burden of mental illness. *The Lancet Psychiatry*, *3*(2), 171-178.

[5] Anand, R., Sindhwani, N., & Saini, A. (2021). Emerging Technologies for COVID-19. *Enabling Healthcare 4.0 for Pandemics: A Roadmap Using AI, Machine Learning, IoT and Cognitive Technologies*, 163-188.

[6] Jain, S., Sindhwani, N., Anand, R., & Kannan, R. (2022). COVID Detection Using Chest X-Ray and Transfer Learning. In *International Conference on Intelligent Systems Design and Applications* (pp. 933-943). Springer, Cham.

[7] Khongsai, L., Anal, T. S., AS, R., Shah, M., & Pandey, D. (2021). Combating the spread of COVID-19 through community participation. *Global Social Welfare*, *8*(2), 127-132.

[8] Ayenew, B., Pandey, D., Yitayew, M., Etana, D., Binay Kumar, P., & Verma, N. (2020). Risk for surge maternal mortality and morbidity during the ongoing corona virus pandemic. *Med. Life Clin.*, *2*(1), 1012.

[9] Ciotti, M., Ciccozzi, M., Terrinoni, A., Jiang, W. C., Wang, C. B., & Bernardini, S. (2020). The COVID-19 pandemic. *Critical reviews in clinical laboratory sciences*, *57*(6), 365-388.

[10] World Health Organization. (2021). COVID-19 weekly epidemiological update, 9 March 2021.

[11] Rather, R. A., Islam, T., Rehman, I. U., & Pandey, D. (2021). Development of vaccine against coronavirus disease 2019 (COVID-19) In India. *Asian Journal of Advances in Medical Science*, 13-21.

[12] Zargar, S. A., Islam, T., Rehman, I. U., & Pandey, D. (2021). Use of Cluster Analysis To Monitor Novel Corona Virus (COVID-19) Infections In India. *Asian Journal of Advances in Medical Science*, 1-7.

[13] Meivel, S., Sindhwani, N., Anand, R., Pandey, D., Alnuaim, A. A., Altheneyan, A. S., ... & Lelisho, M. E. (2022). Mask Detection and Social Distance Identification Using Internet of Things and Faster R-CNN Algorithm. *Computational Intelligence and Neuroscience, 2022.*

[14] Bueno-Notivol, J., Gracia-García, P., Olaya, B., Lasheras, I., López-Antón, R., & Santabárbara, J. (2021). Prevalence of depression during the COVID-19 outbreak: A meta-analysis of community-based studies. *International journal of clinical and health psychology*, *21*(1), 100196.

[15] Ettman, C. K., Abdalla, S. M., Cohen, G. H., Sampson, L., Vivier, P. M., & Galea, S. (2020). Prevalence of depression symptoms in US adults before and during the COVID-19 pandemic. *JAMA network open*, *3*(9), e2019686-e2019686.

[16] Economou, M., Souliotis, K., Peppou, L. E., Agapidaki, I., Tzavara, C., & Stefanis, C. N. (2018). Major depression in Cyprus amid financial crisis: prevalence and correlates. *International Journal of Culture and Mental Health*, *11*(3), 255-267.

[17] Solomou, I., & Constantinidou, F. (2020). Prevalence and predictors of anxiety and depression symptoms during the COVID-19 pandemic and compliance with precautionary measures: Age and sex matter. *International journal of environmental research and public health*, *17*(14), 4924.

[18] Knuth, B. S., Silva, R. A. D., Oses, J. P., Radtke, V. A., Cocco, R. A., & Jansen, K. (2015). Mental disorders among health workers in Brazil. *Ciência & Saúde Coletiva*, *20*, 2481-2488.

[19] Remuzzi, A., & Remuzzi, G. (2020). COVID-19 and Italy: what next?. *The lancet*, *395*(10231), 1225-1228.

[20] Gondi, S., Beckman, A. L., Deveau, N., Raja, A. S., Ranney, M. L., Popkin, R., & He, S. (2020). Personal protective equipment needs in the USA during the COVID-19 pandemic. *The Lancet*, *395*(10237), e90-e91.

[21] Mehrotra, P., Malani, P., & Yadav, P. (2020, May). Personal protective equipment shortages during COVID-19—supply chain–related causes and mitigation strategies. In *JAMA Health Forum* (Vol. 1, No. 5, pp. e200553-e200553). American Medical Association.

[22] [22] Khalil, M. M., Alam, M. M., Arefin, M. K., Chowdhury, M. R., Huq, M. R., Chowdhury, J. A., & Khan, A. M. (2020). Role of personal protective measures in prevention of COVID-19 spread among physicians in Bangladesh: a multicenter cross-sectional comparative study. *SN comprehensive clinical medicine*, *2*(10), 1733-1739.

[23] Woolley, K., Smith, R., & Arumugam, S. (2020). Personal protective equipment (PPE) guidelines, adaptations and lessons during the COVID-19 pandemic. *Ethics, Medicine and Public Health*, *14*, 100546.

[24] Zhou, Y., Wang, W., Sun, Y., Qian, W., Liu, Z., Wang, R., ... & Zhang, X. (2020). The prevalence and risk factors of psychological disturbances of frontline medical staff in china under the COVID-19 epidemic: workload should be concerned. *Journal of affective disorders*, *277*, 510-514.

[25] Restuputri, D. P., Pangesti, A. K., & Garside, A. K. (2019). The measurement of physical workload and mental workload level of medical personnel. *Jurnal Teknik Industri*, *20*(1), 34-44.

[26] Abdelghani, M., El-Gohary, H. M., Fouad, E., & Hassan, M. S. (2020). Addressing the relationship between perceived fear of COVID-19 virus infection and emergence of burnout symptoms in a sample of Egyptian physicians during COVID-19 pandemic: a cross-sectional study. *Middle East Current Psychiatry*, *27*(1), 1-9.

[27] Mamun, M. A., Bodrud-Doza, M., & Griffiths, M. D. (2020). Hospital suicide due to non-treatment by healthcare staff fearing COVID-19 infection in Bangladesh? *Asian journal of psychiatry*, *54*, 102295.

[28] Urooj, U., Ansari, A., Siraj, A., Khan, S., & Tariq, H. (2020). Expectations, fears and perceptions of doctors during COVID-19 pandemic. *Pakistan Journal of Medical Sciences*, *36*(COVID19-S4), S37.

[29] Elawad, M. E., Elmalik, A. B., Hassan, D. M. A., Mohammed, F. S. I., Idris, O. H. K., Abdulaziz, N. S. O., & Ali, I. A. (2021). COVID-19 clinical presentations: the modern mimic of other conditions. *South Sudan Medical Journal*, *14*(1), 5-7.

[30] Yufika, A., Pratama, R., Anwar, S., Winardi, W., Librianty, N., Prashanti, N. A. P., & Harapan, H. (2021). Stigma against COVID-19 among health care workers in Indonesia. *Disaster Medicine and Public Health Preparedness*, 1-12.

[31] Malik, S., Ullah, I., Irfan, M., Ahorsu, D. K., Lin, C. Y., Pakpour, A. H., & Minhas, R. (2021). Fear of COVID-19 and workplace phobia among Pakistani doctors: A survey study. *BMC Public Health*, *21*(1), 1-9.

[32] Bitan, D. T., Grossman-Giron, A., Bloch, Y., Mayer, Y., Shiffman, N., & Mendlovic, S. (2020). Fear of COVID-19 scale: Psychometric characteristics, reliability and validity in the Israeli population. *Psychiatry Research*, *289*, 113100.

[33] Mal, P. R., Suneel, P., & Shomeeta, P. (2020). Social distancing: A non-pharmacological intervention for COVID-19. *J. Pak. Med. Assoc.*, *70*(5^ s3).

[34] Galea, S., Merchant, R. M., & Lurie, N. (2020). The mental health consequences of COVID-19 and physical distancing: the need for prevention and early intervention. *JAMA internal medicine*, *180*(6), 817-818.

[35] Brooke, J., & Jackson, D. (2020). Older people and COVID-19 isolation, risk and ageism. *Journal of clinical nursing*.

[36] Razai, M. S., Oakeshott, P., Kankam, H., Galea, S., & Stokes-Lampard, H. (2020). Mitigating the psychological effects of social isolation during the COVID-19 pandemic. *BMJ*, *369*.

[37] Hummel, S., Oetjen, N., Du, J., Posenato, E., de Almeida, R. M. R., Losada, R., ... & Schultz, J. H. (2021). Mental health among medical professionals during the COVID-19 pandemic in eight european countries: Cross-sectional survey study. *Journal of medical Internet research*, *23*(1), e24983.

[38] Chatterjee, S. S., Bhattacharyya, R., Bhattacharyya, S., Gupta, S., Das, S., & Banerjee, B. B. (2020). Attitude, practice, behavior, and mental health impact of COVID-19 on doctors. *Indian Journal of Psychiatry, 62*(3), 257.

[39] Parthasarathy, R., Jaisoorya, T. S., Thennarasu, K., & Murthy, P. (2021). Mental health issues among health care workers during the COVID-19 pandemic–a study from India. *Asian Journal of Psychiatry, 58*, 102626.

[40] Ning, X., Yu, F., Huang, Q., Li, X., Luo, Y., Huang, Q., & Chen, C. (2020). The mental health of neurological doctors and nurses in Hunan Province, China during the initial stages of the COVID-19 outbreak. *BMC psychiatry, 20*(1), 1-9.

[41] Amin, F., Sharif, S., Saeed, R., Durrani, N., & Jilani, D. (2020). COVID-19 pandemic-knowledge, perception, anxiety and depression among frontline doctors of Pakistan. *BMC psychiatry, 20*(1), 1-9.

[42] Greenberg, N., Weston, D., Hall, C., Caulfield, T., Williamson, V., & Fong, K. (2021). Mental health of staff working in intensive care during COVID-19. *Occupational Medicine, 71*(2), 62-67.

[43] Uvais, N. A., Shihabudheen, P., & Hafi, N. B. (2020). Perceived stress and stigma among doctors working in COVID-19-designated hospitals in India. *The primary care companion for CNS disorders, 22*(4), 26178.

[44] Di Tella, M., Romeo, A., Benfante, A., & Castelli, L. (2020). Mental health of healthcare workers during the COVID-19 pandemic in Italy. *Journal of evaluation in clinical practice, 26*(6), 1583-1587.

[45] Blekas, A., Voitsidis, P., Athanasiadou, M., Parlapani, E., Chatzigeorgiou, A. F., Skoupra, M., & Diakogiannis, I. (2020). COVID-19: PTSD symptoms in Greek health care professionals. *Psychological Trauma: Theory, Research, Practice, and Policy, 12*(7), 812.

[46] Chatzittofis, A., Karanikola, M., Michailidou, K., & Constantinidou, A. (2021). Impact of the COVID-19 Pandemic on the Mental Health of Healthcare Workers. *International Journal of Environmental Research and Public Health, 18*(4), 1435.

[47] Gesi, C., Carmassi, C., Cerveri, G., Carpita, B., Cremone, I. M., & Dell'Osso, L. (2020). Complicated grief: what to expect after the coronavirus pandemic. *Frontiers in psychiatry, 11*, 489.

[48] Anderson, W. G., Arnold, R. M., Angus, D. C., & Bryce, C. L. (2008). Posttraumatic stress and complicated grief in family members of patients in the intensive care unit. *Journal of general internal medicine, 23*(11), 1871-1876.

[49] Conti, C., Fontanesi, L., Lanzara, R., Rosa, I., & Porcelli, P. (2020). Fragile heroes. The psychological impact of the COVID-19 pandemic on health-care workers in Italy. *PLoS One, 15*(11), e0242538.

[50] John, A., Eyles, E., Webb, R. T., Okolie, C., Schmidt, L., Arensman, E., & Gunnell, D. (2020). The impact of the COVID-19 pandemic on self-harm and suicidal behaviour: update of living systematic review. *F1000Research, 9.*

[51] Laboe, C. W., Jain, A., Bodicherla, K. P., & Pathak, M. (2021). Physician suicide in the era of the COVID-19 pandemic. *Cureus, 13*(11).

[52] Brahmi, N., Singh, P., Sohal, M., & Sawhney, R. S. (2020). Psychological trauma among the healthcare professionals dealing with COVID-19. *Asian Journal of Psychiatry, 54*, 102241.

[53] Sharma, G. D., Ghura, A. S., Mahendru, M., Erkut, B., Kaur, T., & Bedi, D. (2020). Panic during COVID-19 pandemic! A qualitative investigation into the psychosocial experiences of a sample of Indian people. *Frontiers in psychology*, *11*, 2494.
[54] Franck, E., Goossens, E., Haegdorens, F., Geuens, N., Portzky, M., Tytens, T. & Van Bogaert, P. (2021). Role of resilience in healthcare workers' distress and somatization during the COVID-19 pandemic: A cross-sectional study across Flanders, Belgium. *Nursing open*.
[55] Gulati, G., & Kelly, B. D. (2020). Physician suicide and the COVID-19 pandemic. *Occupational Medicine*, *70*(7), 514-514.
[56] Jahan, I., Ullah, I., Griffiths, M. D., & Mamun, M. A. (2021). COVID-19 suicide and its causative factors among the healthcare professionals: Case study evidence from press reports. *Perspectives in psychiatric care*.
[57] Bakshi, G., Shukla, R., Yadav, V., Dahiya, A., Anand, R., Sindhwani, N., & Singh, H. (2021). An Optimized Approach for Feature Extraction in Multi-Relational Statistical Learning. *Journal of Scientific and Industrial Research (JSIR)*, *80*(6), 537-542.
[58] Shukla, R., Dubey, G., Malik, P., Sindhwani, N., Anand, R., Dahiya, A., & Yadav, V. (2021). Detecting crop health using machine learning techniques in smart agriculture system. *Journal of Scientific and Industrial Research (JSIR)*, *80*(08), 699-706.
[59] Sindhwani, N., Anand, R., Meivel, S., Shukla, R., Yadav, M. P., & Yadav, V. (2021). Performance Analysis of Deep Neural Networks Using Computer Vision. *machine learning*, *15*, 17.
[60] World Health Organization. (2020). Mental health and psychosocial considerations during the COVID-19 outbreak, 18 March 2020.
[61] Liu, K., Chen, Y., Wu, D., Lin, R., Wang, Z., & Pan, L. (2020). Effects of progressive muscle relaxation on anxiety and sleep quality in patients with COVID-19. *Complementary therapies in clinical practice*, *39*, 101132.
[62] Greenberg, N., Docherty, M., Gnanapragasam, S., & Wessely, S. (2020). Managing mental health challenges faced by healthcare workers during COVID-19 pandemic. *BMJ*, *368*.
[63] Walton, M., Murray, E., & Christian, M. D. (2020). Mental health care for medical staff and affiliated healthcare workers during the COVID-19 pandemic. *European Heart Journal: Acute Cardiovascular Care*, *9*(3), 241-247.

Chapter 10

Environmental Behavior Analysis and Monitoring with IoT for COVID-19

Harinder Singh[1], Y. Prasanna Kumar[2], Shahanawaj Ahamad[3], Sushma Jaiswal[4], M. K. Dharani[5], R. D. Padmavathy[6] and Ankur Gupta[7,*]

[1]Department of CS & IT, Sant Baba Attar Singh Khalsa College, Sandaur, Punjab, India
[2]Department of Mining Engineering, BuleHora University, BuleHora, Ethiopia
[3]College of Computer Science and Engineering, University of Hail, Hail City, Saudi Arabia
[4]Department of Computer Science & Information Technology, Guru Ghasidas Vishwavidyalaya, Bilaspur, Chhattisgarh, India
[5]Department of Computer Science and Engineering, Kongu Engineering College, Erode, Tamil Nadu, India
[6]Mathematics Education, Department of Education, Tezpur University (A Central University), Assam, India
[7]Department of Computer Science and Engineering, Vaish College of Engineering, Rohtak, Haryana, India

Abstract

Environmental monitoring and impact assessment reports are created by compiling and analyzing data. Ensuring that an organization's operations do not harm human health or the natural environment is the primary goal of environmental monitoring. To conserve the earth, safeguard public health, and minimize expenses, environmental monitoring systems are becoming increasingly important. When discussing IoT-enabled

---

[*] Corresponding Author's Email: ankurdujana@gmail.com.

In: An Interdisciplinary Approach in the Post-COVID-19 Pandemic Era
Editors: D. Pandey, B. Kumar Pandey, N. Sindhwani et al.
ISBN: 979-8-88697-052-4

environmental monitoring, waste management, air pollution, and extreme weather typically come up. Infectious illness is induced by a newly discovered coronavirus (COVID-19). A mild to severe respiratory sickness is the most common symptom of infection with COVID-19 and does not necessitate any special treatment for most patients. An increased risk of serious sickness is seen in the elderly, as well as individuals who have pre-existing medical conditions. COVID-19, the virus that causes and transmits the disease, can be prevented and slowed down by being educated about it. Using an alcohol-based hand sanitizer or washing your hands often is a good way to keep yourself and others safe from infection. Different people are affected by COVID-19 in various ways. A viral infection usually results in mild to moderate symptoms, and the majority of individuals will be able to heal at home without the need for medical attention. Sensors are used to monitor and record every aspect of the environment in their respective settings. The most commonly utilized sensors include proximity sensors, temperature sensors, smoke sensors, etc. With an ever-increasing population, increasing industrial activity, and rising energy consumption, it is imperative that the environmental monitoring process be made more accurate and efficient by using cutting-edge, automated monitoring applications and devices. It will be a useful tool for monitoring climate change. People, energy, goods, and information will move more efficiently as a result. In addition, it will continue to adapt as scientists and leaders come to new agreements on the actions, we must take to safeguard ourselves from rising temperatures and hazardous climate circumstances.

Keywords: environmental monitoring, IoT, sensors, smart environmental monitoring, applications and challenges of SEM, COVID-19

10.1. Introduction

There is no longer any mystery about the Internet of Things. It's a technology that has been quietly changing our future for some time now. As a result of human ingenuity and a desire to live a more comfortable and interconnected lifestyle, the Internet of Things was born. We decided to make gadgets smart and to take care of things that pull efficiency. An excellent outcome has been reached with this step. Internet of Things (IoT) monitoring of the environment is essential for conserving energy, water, and other natural resources, all of which are being depleted at an alarming rate. Samples are collected, examined, and analytical instruments are used in standard environmental monitoring approaches. To accomplish this, there are two options. First, the material is gathered and evaluated in a laboratory by hand. The second method is instrumental, in which the number of contaminants in a sample is automatically assessed. There is no need for manual data entry or analysis while using instrumentation. The sample must be pre-treated in manual procedures before sedimentation, isolation, and other operations may be performed on it. When discussing IoT-enabled environmental monitoring, waste management, air pollution, and extreme weather typically come up. IoT-based devices are frequently used for disease prediction and symptom analysis in the healthcare sector [1]. To enhance IoT Security, Privacy, Safety, and Ethics [2], a Blockchain framework [3] came into existence where IoT has been applicable in industry-oriented applications [4]. Researchers did work to provide a Multi-level trust-based intelligence schema to secure IoT against security threats using cryptographic authentication [5]. During the further evolution of IoT concept of machine learning [6] is also introduced to provide energy-efficient routing [7] IoT-based smart healthcare monitoring systems [8] gained popularity. The famous application of IoT makes use of the unsupervised intelligent system with Grey Wolf optimization for IoT [9]. These applications might be intrusion detection Systems [10], and DDoS defense mechanisms [11].

But it has been observed that there are several opportunities and challenges [12] in the implication of IoT. IoT is dependent on devices, gateways, operating systems, middleware, and data communication [13]. Further researches make use of blockchain-based authentication mechanism in IoT-based systems during transaction processing [14-16]. Internet of Things (IoT) applications have been used for elderly care [17], signal processing [18], policing processes [19], and big data processing [20].

10.1.1. Environmental Monitoring

Environmental monitoring refers to the tools and procedures used to observe, describe, and define environmental parameters to determine exactly the environmental consequences of a certain activity. Environmental monitoring and impact assessment reports are created by compiling and analyzing data [21]. Ensuring that an organization's operations do not harm human health or the natural environment is the primary goal of environmental monitoring. As the world's population, industry, and energy consumption continue to grow, it is more important than ever to create environmental monitoring apps and devices that are advanced, automated, and easy to use.

10.1.2. Rationale behind Environment Monitoring

Understanding and reducing consequences, adhering to environmental laws, and protecting the environment and human health are the primary objectives. Environmental impact assessments (ESAs), which investigate the potential negative effects and consequences of a proposed project on the surrounding environment, must be conducted using state-of-the-art environmental monitoring equipment. As a result, environmental monitoring is a prerequisite for the start of each firm. Environmental monitoring metrics provide insight into how the environment is responding to the activities taking place around it. Environmental monitoring systems play a critical role in reducing environmental impact, saving the planet, limiting public exposure to toxins, and reducing expenses at a time when we are using more energy.

10.1.3. Types of Environmental Monitoring

The major 4 types of environmental monitoring including biodiversity, air quality, water quality, and noise quality are as:-

10.1.3.1. Environmental Monitoring of Air

Pollutants in the air impair both human health and the ecosystem [21]. Businesses and organizations must abide by national air quality rules due to the importance of air quality to human health. Real-time monitoring of air quality can be accomplished by connecting sensors in a chosen area to a piece of software [22]. For those who prefer the old-fashioned approach, stationary air particle collectors can be used to collect samples of the atmosphere and then analyze them afterward.

10.1.3.2. Environmental Monitoring of Water

As water is fundamental to life on Earth, all bodies of water must be monitored. Clean, safe, and hygienic water is a necessity. Marine life and biodiversity can flourish only if water bodies are clean. Connecting sensors in water to software allows for real-time water measurement. Dissolved oxygen, turbidity, and bio indicators like nitrates and pH are examples of thresholds that can be communicated by these sensors. It is also possible to collect water samples by hand and send them to a laboratory.

10.1.3.3. Environmental Monitoring of Noise

In both the water and on land, noise may be quantified. Whales and dolphins have the same problem, as well. For instance, whales may be unable to communicate due to noise pollution, putting them at risk. On the ground, noise can harm animals and human well-being. So that organizations know exactly how much noise they can make, noise level thresholds are established in advance. In the ocean, noise monitoring safeguards marine life, while on land, communities near industrial areas like ports and airports are safeguarded by noise monitoring.

10.1.3.4. Environmental Monitoring of Biodiversity

The diversity of life on Earth is referred to as biodiversity. We're seeing fewer species of plants and animals due to climate change. Biomonitoring reveals how our activities affect the ecosystem directly because of biodiversity responses. Over time, a variety of animals and plants are studied to see how their environments are being affected by human activity.

10.1.4. Right to Clean Environment

Noise reduction, environmental monitoring, and biomonitoring are all vital for human survival. To stay in business and launch new initiatives, companies must keep an eye on their long- and short-term impact on the environment because of a growing number of government laws. Public support can also be gained as a result.

10.1.5. IoT

The Internet of Things (IoT) was created to allow intelligent devices to communicate with one another as the next big disruptive technology industry [23]. Through the integration of diverse technologies from many applications, it connects digital and physical transactions. The term "smart" now encompasses a wide range of cutting-edge devices with some level of artificial intelligence [24], not just those found in the average home. System automation, analysis, and integration can all be improved using IoT technologies [23]. They increase the scope and precision of these areas. Sensors, networks, and robots are all part of the IoT, which uses both established and new technologies. Software advancements, reducing hardware prices, and modern views toward technology all contribute to IoT's success. There is a huge shift in the delivery of products, goods, and services as a result of its new and advanced elements. Internet of Things (IoT) monitoring of the environment is essential for conserving energy, water, and other natural resources, all of which are being depleted at an alarming rate. Samples are collected, examined, and analytical instruments are used in standard environmental monitoring approaches. To accomplish this, there are two options. First, the material is gathered and evaluated in a laboratory by hand. The second method is instrumental, in which the number of contaminants in a sample is automatically assessed. There is no need for manual data entry or analysis while using instrumentation. The sample must be pre-treated in manual procedures before sedimentation, isolation, and other operations may be performed on it. When discussing IoT-enabled environmental monitoring, waste management, air pollution, and extreme weather typically come up.

10.1.5.1. Key Features of IoT

AI, connection, sensors, active participation, and small device utilization are the most significant aspects of IoT [25]. The following is a quick rundown of these features:

a) Everything may become "smart" thanks to the power of data collecting, AI algorithms, & networks, which is what IoT is all about. This might be as simple as installing sensors in your fridge and cabinets to alert you when you're running low on milk and pulses and then placing a command with your local grocery store.
b) IoT networks are no longer confined to a few big suppliers, thanks to new enabling technologies for networking, such as Wi-Fi and Bluetooth. These micro-networks are created by IoT between its system components.
c) IoT would be nothing if it didn't have sensors. IoT becomes an active system capable of real-world assimilation with help of these defining instruments.
d) Connected technology is increasingly being used passively, rather than actively. There is a new way to actively engage with content, products, and services via IoT.
e) Small Devices - As expected, devices have shrunk in size, cost, and power. The precision, scalability, and flexibility of IoT are made possible by the use of custom-built tiny devices.

10.1.5.2. Architecture Behind IoT Monitoring

This section of the chapter presents information on the working of IoT. Figure 10.1 is a diagrammatic representation of the architecture behind the internet of things technology.

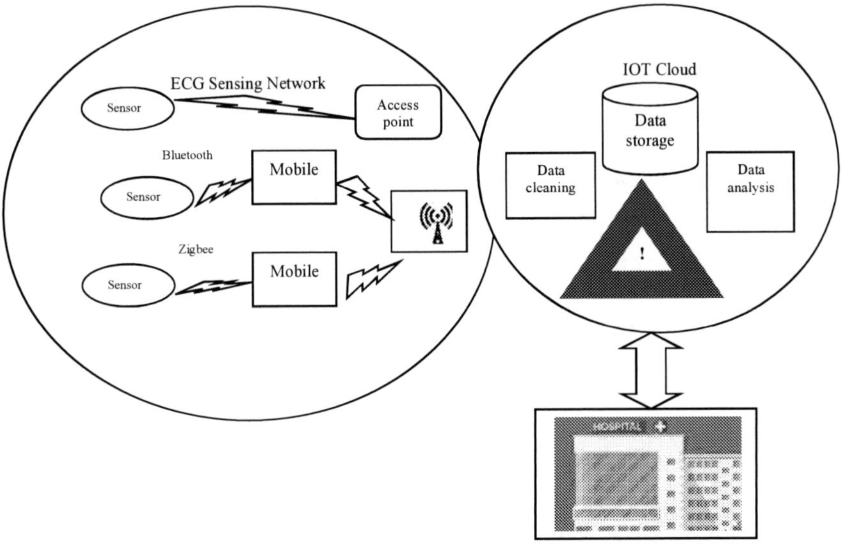

Figure 10.1. Working of IoT.

As shown, the smart environment sensing devices create data or trigger information that is sent to the database via a smart meter. Then preprocessing and classification of data take place. After that data clustering is performed. Based on that data, activity is predicted. This provides us with a refined database. From the selected features, the alert can be generated.

10.1.6. IoT and Its Hardware

Devices, Sensors, servers, and bridge devices or routing for a remote dashboard are all components of IoT systems [26]. To support specified goals and activities, these devices control critical duties & functions.

1. Sensors may be the most significant piece of hardware in IoT. To put it another way, these gadgets are made up of several different components. Signal processing, Wi-Fi, Zig-Bee, Bluetooth, radio transceiver, duplexer, and BAW all play a role in RF modules' ability to communicate. Assorted active and passive sensors are used in the sensing module. Microflow sensors, temperature sensors, magnetometers, proximity sensors, and gyroscopes are some of the measurement devices utilized in the Internet of Things (IoT). Other IoT measurement devices include accelerometers and gyros.
2. Gadgets worn on the feet, head, neck, arms, and chest are all examples of wearable electronic devices. As a component of an IoT system, smartwatches not only keep us connected but also provide them access we need to boost our productivity. In the same way, they open up new possibilities for productivity.
3. The command center and remotes of IoT are still the PC, tablet, and cellphone. Using the desktop gives you the most control over your computer's operating system and its configuration options. The tablet may be used as a remote control as well as a method to access the system's major functionalities like a desktop. Some basic settings may be changed on the phone, and it also has remote access capabilities.
4. Routers and switches, which are standard components of any network, are also important components that must be linked.

10.1.7. IoT and Its Software

There are several ways in which the IoT software solves its activity domains and primary networking. They gather data from sensors and devices and perform real-time process and analytics extensions in IoT networks using individual and master apps [27]. They make use of the integration of important business systems to perform related activities.

a) Light data filtering, data security, and aggregation are all handled by this software's sensing and measuring functions. To link sensors to real-time, machine-to-machine networks, it makes use of certain protocols. Data is collected from different devices and distributed in line with the parameters that have been established. It can also be used to distribute data backward. At some point, the system will send all of the data it has collected to a central server.

b) The software enabling device integration ties together (creates dependent connections) all of the system's components to form the IoT system's body. As a result, gadgets can operate together seamlessly and reliably. The Internet of Things (IoT) would not exist without these apps, which constitute the network's defining software technology. It is their job to coordinate the numerous protocols, applications, and restrictions of each device to make communications possible.
c) In real-time analytics, data from a variety of sources are transformed into useful actions or patterns that can be analyzed by a human. To conduct automation-related activities/give the industry with data it requires; they evaluate data depending on various settings and designs.
d) Existing systems and software may be made more useful by incorporating these applications and processes. Some preset devices, such as mobile devices or technical equipment, are integrated into the system. It facilitates increased output and more precise data collecting.

10.1.8. Need of the IoT for Environment Monitoring

We can't just start employing sensors for regulatory purposes when we get into the nitty-gritty of environmental monitoring. We can apply AI and ML methods [28], among others, if we have data on water and air. Water quality, radiation, and toxic substances can all be measured with environmental sensors. We also need industrial IoT (IIoT) technologies to ensure worker safety because some businesses produce harmful gases like sulfur, methane, and sulphur's compounds, which are harmful to human health. We can keep our safety record spotless by collecting sensor data. Sensors can be used in otherwise inaccessible places. Research on environmental sensors has been ongoing since 2012 all around the world. The results of a few audits have already been made public. A ZigBee-based wireless sensor network has been used to monitor outdoor air quality. However, because indoor air is more contaminated than outdoor air, a system designed for outdoor air may not be appropriate for indoor use. To keep track of airborne particulates, scientists have created a new air quality monitoring system (PM). There are three levels of categorization in the system: PM 10, PM 2.5, and PM 1. When PM enters the lungs, it can cause a variety of health problems. Monitoring is focused on PM 1 but also on PM 10 and 2.5 in particular. PM 2.5 concentrations in cities can now be monitored using mobile sensing systems that have been designed and tested. Papers in the scientific literature describe low-cost, portable monitoring systems that can measure a wide range of environmental parameters such as CO2, CO illuminance, and sound levels.

10.1.9. A Comparison between Traditional Internet and IoT

The details provided in Table 10.1 shed light on the key changes IoT has brought to the traditional internet.

Table 10.1. Key Changes in IoT and Tradition Internet [1]

Topic	Traditional Internet	IoT
Content creation	By human	By machines
Mechanism of combining content	By explicitly defined links	By explicitly defined operations
Value	Answers the query	Action and timely info
Connection type	Point-on-point and multipoint	Multipoint
Digital data	Readily provided	Not generated until augmented or manipulated
Data formats	Homogeneous	Heterogenous
Composition	PC, server, and smartphones	RFID and WSN nodes
Content Consumption	By request	By pushing info and triggering actions

10.1.10. Application Areas of IoT other than Smart Environmental Monitoring

The use case of IoT is very high and multipurpose right from smart TV in-home to whole smart home by using home-automation [29]. Also, its use is from domestic use to a wide array of industrial activities. Application areas are shown in Figure10.2.

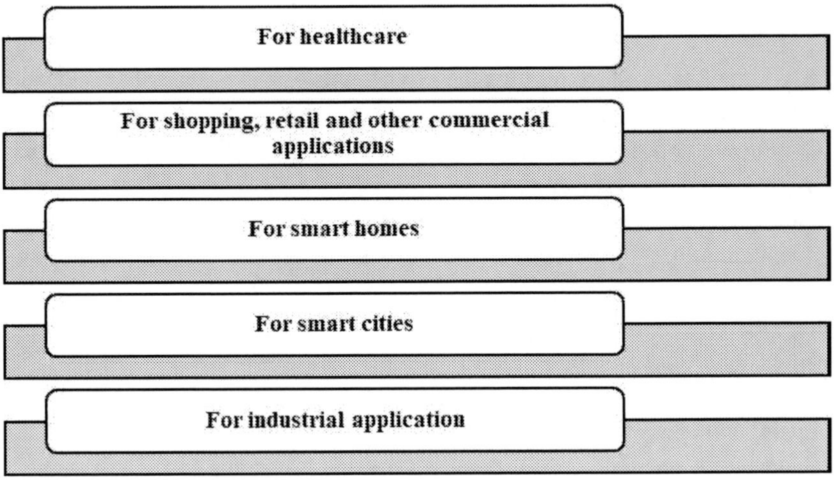

Figure 10.2. Application areas of IoT other than SEM.

Healthcare: IoT helps in patient monitoring [1]. Also, it can aid in keeping track of people's health in old-age homes. Smart wearables have shown to be very effective, recently when an apple watch helped a cardiac arrest patient by informing the nearby healthcare centers.

256 organizations and startups are included in the IoT Analytics Smart Home company database [29]. There are more firms involved in smart home technology than any other IoT application. Smart Home firms have raised more than $2.5 billion in financing so far. Several household brands like Nest and Alert-Me are included, as well as global firms like Philips and Haier.

From traffic control and water distribution to trash management, urban security, and environmental monitoring, a smart city encompasses a broad range of applications. Because many Smart City solutions claim to ease the genuine problems of people who live in cities today, it has become more popular. Smart City IoT solutions minimize traffic congestion, noise, and pollution, all while increasing public safety.

Industrial application: There are various industrial applications from supply chain management to predicting maintenance of machines, quality checks, and aiding in inventory management. SCM with IoT is becoming a reality. Artificial intelligence is also being supplemented with both of them to get a multiplier effect of technology.

10.1.11. Application of IoT in Smart Environmental Monitoring

The IoT may be used to monitor the environment in a variety of ways, from tracking severe weather and water quality to safeguarding endangered species and commercial farming. Sensors are used to monitor and record every aspect of the environment in these settings. Current air and water safety monitoring technology is improved by IoT since it requires less human labor, allows for frequent sampling, broadens the scope of sample and monitoring, and allows for more advanced testing on-site. As a result, we're able to avoid major contamination and the ensuing catastrophes. It's difficult to accurately monitor weather conditions in commercial farming because it relies on humans to do it. It also has a small amount of automation. System functions, agricultural analysis, and monitoring can be automated with IoT. Crop, soil, and environmental changes are all detected by these systems [30]. Analyzing a wide range of data sets improves the efficiency of established procedures. Health problems (such as E. coli) can also be prevented thanks to these devices.

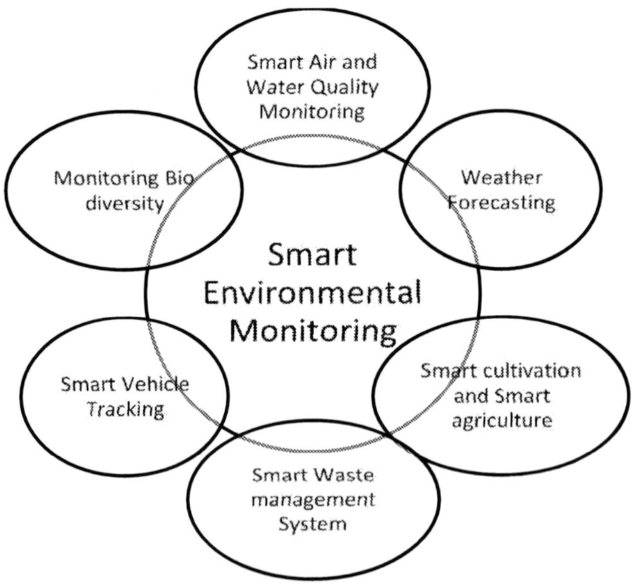

Figure 10.3. Depicts application areas of IoT in smart environmental monitoring.

1. *Monitoring Air and Water quality*: Manual labor and modern instruments are used in conjunction with lab processing to assess air and water quality. With IoT, human labor is reduced, sampling can be done more frequently, sampling and monitoring can be expanded, on-site testing is made possible, and reaction activities are tied to detecting systems. As a result, we're able to avoid major contamination and the ensuing catastrophes.
2. *Extreme Weather*: The use of broad equipment, like radar and satellites, instead of more granular solutions, limits the ability to detect extreme weather. Because they use less powerful technology, they can't attain the same level of precision with their instruments for minor details. More precise data, increased precision, and increased adaptability are all promised by new IoT advancements. High levels of detail and adaptability in terms of range, instrument type, and deployment are essential to effective forecasting. Preventing loss of life or property is made possible by the ability to recognize and respond quickly in an emergency.
3. *Commercial Farming*: For a long time, modern technology and biotechnology have been used on commercial farms, but the Internet of Things (IoT) provides new opportunities. Many aspects of commercial farming lack precision and necessitate the use of human labor. It also has a small amount of automation. System functions, agricultural analysis, and monitoring can be automated with IoT. Crop, soil, and environmental changes are all detected by these systems. Analyzing a wide range of data sets improves the efficiency of established procedures. Health problems (such as E. coli) can also be prevented thanks to these devices.
4. *Waste Management*: Large cities face two major challenges when it comes to waste management: rising service costs and limited space for the city's growing rubbish piles. Information and communications technology solutions will need to be more widely used in this area to save money and take advantage of environmental advantages. It is possible to lower the cost of waste collection and increase recycling quality by using intelligent waste containers, such as those that measure the weight of the waste they contain. The Internet of Things (IoT) will connect these smart waste containers to a control center, where optimization software will process the data and figure out the best path for the collector truck to take to effectively utilize these smart waste management services.

5. *Vehicle Tracking*: An intelligent display system helps drivers locate the optimal route for parking in the city by using road sensors and clever display systems; there are numerous advantages to using this service, such as less CO emissions from the car, and fewer traffic jams, and overall happier residents. The parking facility can be immediately integrated into the IoT infrastructure. The use of communication technologies like Near Field Communication (NFC) or Radio Frequency Identifiers (RFID), as we discussed earlier, also enables us to better serve residents who can benefit from the parking spaces reserved for residents or disabled people while also quickly spotting any infractions.
6. *Biodiversity*: It's wonderful to see how IoT is being used to combat climate change and conserve biodiversity, stop deforestation and poaching, and collect data. Many IoT projects are being developed by the Rainforest Connection to avoid further deforestation, such as halting illicit logging in its tracks. To monitor and detect illicit logging and poaching (more on that below), they've built sensors out of old cell phones that can be mounted to trees. For example, a chainsaw or a heavy vehicle in an illegal region can be detected by sensors that listen to these sounds. This allows authorities to stop illegal logging in real-time. Since bees are essential to ecosystem diversity, their recent fall has had a significant impact on agriculture and farming. Beekeepers may maintain tabs on the health of their hives by inserting a battery-operated sensor into each one. Motion-sensing cameras have been used by wildlife preservation organizations to identify uncommon species and the prevalence of illicit hunting and poaching in vast, natural areas that are difficult to monitor.

10.1.12. Glance at Sensors

Electrochemical sensors and metal-oxide sensors are two types of sensors. Depending on the needs of the company, these sensors can be used. In both types of sensors, there are benefits and drawbacks [31]. Metal-oxide sensors, on the other hand, are the focus of environmental monitoring research. To monitor airborne fine particulates in real-time, a micro sensor-based air quality monitoring system has been created in the same way. It has already been put through its paces and is ready for prime time. We can't build a protocol-based system and expect it to work in every case, because the world is so diverse. As a result, a multi-protocol system is required. Understanding the interference of pollutants, such as ozone, nitrogen dioxide, or nitrogen oxide particles, is critical. The science behind this interference, the methods by which data is collected, and the potential causes of data skewness must thus be investigated and clarified. It is only at this point that a sensor may be fully defined and developed to its full potential.

In the real world, the most common sensors are:

a) *Temperature sensors*: They are used in household appliances such as refrigerators, ACs, and thermostats as well as in industries such as manufacturing where machines are required to operate at certain temperatures. Temperature sensors can provide real-time information about the temperature of machinery and the natural environment, helping carry out operations at an optimum temperature.
b) *Proximity sensors*: Proximity sensors are used to detect nearby objects. These sensors are largely used in the autonomous vehicles sector and warehouses by autonomous robots. Proximity sensors emit electromagnetic radiation, for example infrared, and detect changes in the return signals. Some of the types of proximity sensors include Inductive sensors that detect metallic objects in their vicinity using electromagnetic radiation. Capacitive sensors are used to detect metallic as well as nonmetallic objects by a change in the amplitude caused by the entrance of an object in the electrostatic field of electrodes. Lastly, Photo-electric sensors use a beam of light to detect the presence of objects. Since light can travel long distances, these sensors are used for long-range sensing purposes.
c) *Water grade sensors*: Turbidity, pH, oxygen level, TDS, and salinity are some of the factors that can be detected by remote water quality monitoring equipment. With the use of water quality monitoring sensors, an IoT system can keep track of these factors in real-time, allowing for actionable insights and informed decisions to be made.
d) *Gas sensors, which detect air quality conditions*: They are mainly used in oil and gas, manufacturing, and pharmaceutical industries. Gas sensors consist of a sensing material and a heater with free clectrons flowing through them. Oxygen attracts the electrons to the surface of the sensing material, resulting in a low number of electrons freely flowing inside the sensing material. However, if the pure air is replaced by toxic gases such as carbon monoxide or propane, the

electrons held by the oxygen molecules are pushed inside the sensing material that allows the passage of current through the circuit. This triggers the sensor to sound an alarm detecting the presence of harmful gas.

e) *Smoke sensors, which are needed for industrial environmental conditions or smoke-prone places*: Fire and smoke sensors are mounted on a single device so that people receive the alerts related to those hazards in their surroundings.

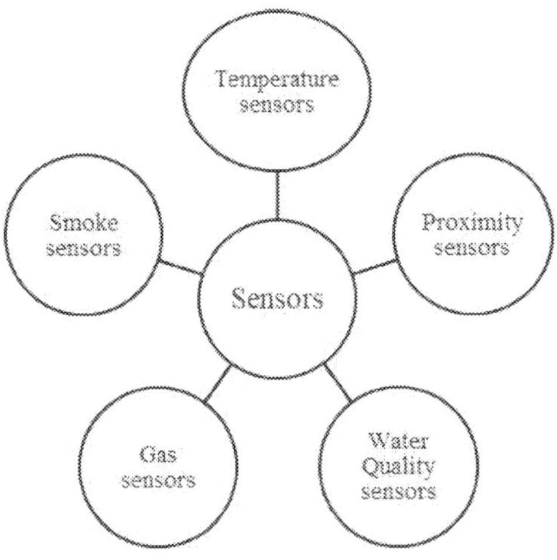

Figure 10.4. Sensors are used for monitoring by IoT technology.

Sensors have been evaluated by the US Environmental Protection Agency (USEPA) using traditional methods to be used for research and IoT applications. The Alphasense OPC N2 sensor measures particulate matter (PM) 10 and 2.5. GRIMM, a certified portable monitor, was used to evaluate this. These sensors haven't received any international recognition as of yet. Rather than USEPA-certified sensors, these are sensors that have been assessed by the agency. This is because new technology is always being developed. Sensors like these are updated every six months with a new model. Gas-phase sensors AQMesh, CairClip, and CitySense can be used to monitor air quality. Projects funded by other countries are being assessed by the US Environmental Protection Agency (EPA). Standard instrumentation methods are being used as a benchmark for the new systems.

This type of pricey regulatory monitor relies on analytical procedures (not sensors). It's dependable, but it's also a bit of a slog. In addition, it must be operated by trained personnel. It has the advantage of having a service life of over ten years. However, it must be re-calibrated every quarter. A low-cost monitor, on the other hand, does not necessitate much training, but its lifespan is limited.

10.1.13. Issues in Deployment of IoT-Based Sensors

When items are synchronized via the Internet using sensors, therefore, following challenges are of crucial importance. Sophisticated interoperability: semantic interoperability; service discovery; service composition; data sharing and cooperation; and autonomous agent identification. While the existing technology delivers a limited amount of information, it is time-consuming and requires a high level of knowledge. The National Environmental Engineering Research Institute is creating a low-cost, easy-to-use, and continuous data-gathering technology that incorporates the Internet of Things. However, a QA/QC approval is required for this type of technology, and there is no one agency in charge of this.

Microprocessors have been the focus of most sensor and system research. Many low-cost components (ranging from $100 to $300) are being used in the creation of the new system at the National Environmental Engineering Research Institute (NEERI). It's also impossible to integrate these components well if you are not a good integrator. Additionally, a single board cannot have more than one sensor. However, if power, cost, and latency can be kept in harmony, the system can be used in the real world. Measurement time, data quality, and budget are all considerations for users, as are sensor features including stability, detection limit, repeatability, and reproducibility. After two or three years, a sensor has reached the end of its useful service life. The sensor's sensitivity, stability, and longevity must be enhanced before it may be used.

The air quality index was developed by the Indian government (AQI). Air quality may be judged using only one number, one color, and one description. It is clear from this that PM 10 is the most harmful of the several pollutants studied. CSIR installed 10 Internet-of-Things (IoT) sensors in Delhi as part of its case study. Low-cost sensors have dominated the market, and just a few sensing elements have been discovered. Sensing elements require additional study. Compared to gas-phase sensors, PM sensors are more commonly available. Sensor applications face two fundamental difficulties: 1), there is a wide range of sensor performance values. 2) Manufacturers are far behind in basic testing.

Installation, upkeep, and data analysis all have to be made more affordable. In the future, we hope for smooth implementation, high data quality, and long-term dependability. To achieve long-term growth through SEM, it is necessary to deal with the major problems of using smart sensors, AI, and WSN. SEM efforts would be bolstered by the cooperation of environmental organizations, regulatory entities, and public awareness. It is possible to improve the quality of the sensory data by preprocessing it using appropriate filters and signal-processing technologies [31].

10.1.14. Challenges in the Path of IoT

IoT has quickly become a significant element of our daily lives, communication, and business practices. Web-enabled devices are transforming our global rights into a larger, more accessible place in which to live. The IoT faces a variety of difficulties. For sake of convenience, here we are providing a taxonomy of challenges (as shown in Figure10.5).

1. Security-related issues [2]
2. Designing related issues
3. Deployment related issues

All of these are discussed ahead in detail.

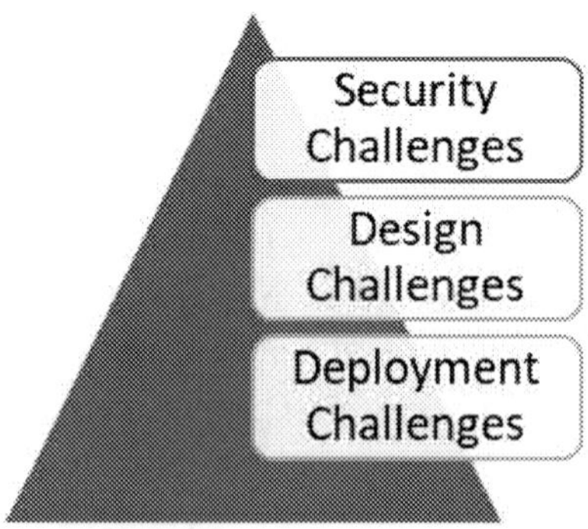

Figure 10.5. Challenges faced by IoT.

10.1.14.1. Security Challenges in IoT

The challenges related to security are discussed here.

- A great way to protect hackers from accessing data, encryption is also one of the most critical challenges in IoT security. The storage and processing capabilities of a typical computer are similar to those of these drives. As a result, the algorithms put in place to keep hackers safe are increasingly vulnerable to assault [32].
- IoT (internet of things) device manufacturers are eager to develop & deploy their gadgets as quickly as they can without giving security too much of a consideration. Insufficient testing and updating. As a result, many of these gadgets and IoT items are vulnerable to hackers and other security breaches.
- Password brute-force attacks and the danger of using a default one – Weak credentials and login details expose nearly all IoT devices to brute force password cracking. A brute force attack can compromise a company's assets as well as its customers' sensitive data if the company's devices are still using factory default credentials.
- IoT As for ransomware, it exploits encryption to prevent users from being able to access their valuable data across a wide range of devices and platforms. Using a computer camera as an example, a hacker can take images. Hackers can extort money from victims by employing malware access points.
- IoT botnet workers can modify data privacy, which might pose a huge risk for an open bit-coin market. Malicious hackers pose a threat to the precise value and development of cryptocurrencies code. It's a priority for block-chain companies to improve security [3]. The app development process is more insecure than the blockchain technology itself.

10.1.14.2. Design Challenges in IoT

The challenges related to designing are discussed ahead.

- The battery life is limited because of packaging and integration issues with small-sized chips that are light and consume little power. If you've been paying attention to the smartphone market, you've probably noticed that the display screen size appears to be unrestricted every year. The rise of 'phablets', which can be as large as tablets, is one example of this. However, larger monitors aren't always just for the sake of convenience; rather, they're expanding to accommodate greater battery capacities. Computers have shrunk in size, but their battery capacity has not.

- In terms of cost, embedded systems aren't as constrained as systems that aren't embedded. Using digital electronic components in the design of IoT devices necessitates better approaches to cost modeling or cost optimization. Additionally, designers must deal with the issue of design time and release the embedded device onto the market promptly.
- Systems must be implemented & developed to be reliable and robust, as well as secure through the use of cryptographic methods and security protocols.

10.1.14.3. Deployment Challenges in IoT

The challenges which create hurdles in IoT implementation are cited here [4]:

- Devices, programs, and cloud platforms all need to be connected to function. The value of connected devices cannot be overstated, especially those that enable quick access to relevant data. But when IoT sensors are needed to monitor process data and supply information, bad connectivity becomes a problem.
- IoT apps must be created with future technology advancements in mind to be cross-platform capable. To create it, hardware and software functions must be balanced. IoT application developers face a problem in ensuring that device and IoT platform drivers deliver optimal performance despite high device rates and fixes.
- Collection and processing of data are critical in IoT development. The processing or usefulness of the stored data is more important in this case. Additionally, development teams must ensure that they plan well for the collection, storage, and processing of data in the environment.
- It is impossible to overcome any of the development issues listed above if the IoT application development team has the right skill set. As long as you have the appropriate people on board, you'll be able to overcome even the most difficult obstacles while building IoT applications.

10.1.15. Green IoT

The IoT is a term used to represent a variety of technologies & study fields that allow physical items to communicate with each other across the globe. Technological enablers such as Nanotechnologies, biometrics, and RFID (radio-frequency identification) are already popular. Bringing the Internet of Things (IoT) into real-world implementations for a wide range of purposes, such as e-health, smart grids, and intelligent transportation are all examples of this. They predict an exhilarating future in which everything is connected in some way, green networks connect us to our physical world Networks of trees. The Internet of Things (IoT) will help reduce emissions and pollution [33]. Making use of environmental protection and monitoring, and reducing the amount of money and energy used.

The amount of information that surrounds us will more than double in the next several years. We will be able to use sensors, gadgets, and “things” to offer green energy via 5G communication, “intelligent” behavior, and help for the management of tasks by users. These new “smart” devices are a big improvement. Context-aware objects will also be able to conduct certain tasks, and performs on their own, requiring new types of green communication from one thing to the next and within objects themselves, where it is possible to optimize power consumption and efforts are made to maximize bandwidth use. As a result of this advancement, it will be relevant to businesses as well as scholars as well as persons [34-37].

10.2. Literature Review

Bala Anand Muthu et al. [1] had considered IoT-based wearable sensors for disease prediction and symptom analysis in the healthcare sector. Food habits and environmental conditions have made it harder for people of excellent health to thrive in today’s world. As a result, if they were to survive, they must raise public knowledge about health issues.

Hany F et al. [2] presented a revolution on the internet that might connect environmental equipment that was over the internet to exchange their data to create novel applications & services that would improve our quality of life. IoTused inexpensive sensors to address, identify, and locate a wide range of devices and things around us.

DeepaPavithran, et al. [3] did research on building a blockchain framework for IoT. In addition to cryptocurrency, blockchain was a very promising technology. Among other things, it's still not obvious how this would operate in IoT networks. More research was needed. For the most part, IoT devices and blockchain technology are both built on ledgers. Numerous possibilities were possible with the Internet of Things (IoT). If the features of the blockchain could be adapted to suit it, as a result, many of the present IoT issues could be addressed.

Fadi Al-Turjman, et al. [4] expressed embedded sensors collecting data from the Industrial Internet of Devices, a rapidly expanding network of networked things. In their system, a variety of things were connected via a diversified communication system that was capable of monitoring and gathering valuable and challenging information. Multimedia and video streaming were likely to play an important role in IIoT-related applications involving Unmanned Aerial Vehicles (UAVs) because of their autonomy, high mobility, and communication and processing power.

Kobra Mabodi, et al. [5] had considered a multi-level trust-based intelligence schema for securing IoT against security threats. A connected, universal, and smart node with autonomous interaction was predicted by the IoT when it comes to services. IoT devices were prime targets for grey whole attacks due to their high computing power, open architecture, and extensive distribution.

Maninder Jeet Kaur, et al. [6] had the potential to change the way they think about globalization in the future. Most businesses throughout the world were likely to be affected by Digital Twin technology, which replicates the physical model for remote monitoring, viewing, and control.

Fatemeh Safari, et al. [7] expressed a priority-based energy-efficient routing method for IoT systems. IoT devices used sensors & monitors to acquire vast amounts of data about their surroundings. One of the numerous difficulties that IoT systems provide was the transmission of data collected by IoT devices to the cloud via relay nodes. Prinergy was proposed in their paper to reduce energy consumption.

Md. Milon Islam, et al. [8] developed Smart Healthcare Monitoring System in IoT Environment. Health care consultations can now move from in-person to telemedicine with ease, thanks to the proliferation of the Internet of Things (IoT) technologies. Patients' fundamental health signals and the room conditions where they were currently located can be monitored in real-time by a smart healthcare system in an IoT environment proposed in their study. The prototype that was created was easy to design and use. Infectious diseases like a novel coronavirus (COVID-19) treatment could benefit greatly from the system. Many lives could be saved as a result of the new system's improvements to the current healthcare system. The system appeared to be somewhat large; it would be reduced in size shortly through good manufacture.

Amaal Al Shorman, et al. [9] considered an unsupervised intelligent system based on a one-class support vector machine. Proliferations of vulnerable IoT devices had resulted in a sharp rise in botnet attacks in recent months. Botnets are a prevalent danger that exploits the lack of basic security mechanisms in IoT devices and could carry out a series of DDoS attacks.

Abhishek Verma, et al. [10], was the most popular study area now. Properties of IoT on one side make it easily relevant to real-life applications, \whereas on the other side expose it to cyber dangers. DoS is one of the most catastrophic attacks against IoT. In their study, they evaluate the potential of employing machine learning classification algorithms for safeguarding IoT against DoS attacks.

Ruchi Vishwakarma, et al. [11] had arisen as a key innovation in the use of wireless media in the real world. A wide variety of smart applications operating on various platforms and virtually everywhere in the globe allow us to exert influence over our immediate environment. An increasing number of malevolent entities were using the Internet of Things (IoT) because it was so widely used. By taking advantage of the Internet of Things (IoT) vulnerabilities resulting from factors such as a lack of resources or inadequate security, these criminals get access to authorized devices.

Mostafa Al-Emran, et al. [12] would be the most difficult platform to implement. IoT and its applications have been the subject of a wide range of review papers that have attempted to assess and synthesize its use. A lot of studies have shown that IoT-based learning frameworks were successful. A shift in their direction contributes to the development of a new educational paradigm. Their new paradigm of learning serves to improve both learning and teaching. IoT adoption and prospective applications on campuses were still rare despite some educational institutions implementing it, especially in underdeveloped countries. Institutional stakeholders must learn how to analyze the demands of students, propose and construct the necessary sensors and software for the Internet of Things applications, and define standards for their effective use. While reviewing studies, researchers found no empirical evidence to back up claims that IoT apps can improve students' learning outcomes.

Sharu Bansa, et al. [13] was playing an increasingly important role in the development of new applications in a wide range of fields, including healthcare, education, smart cities, the home, and agriculture, among others. The IoT ecosystem was surveyed in their article. A wide range of new and innovative M2M technologies was on the horizon, each with its own unique set of capabilities and features to offer users. The goal of their study was to describe the boundaries of elements in a huge heterogeneous IoT ecosystem, from the feet of correct elements forward. The IoT ecosystem may be better understood by looking at all of its components, including the architecture, devices, operating system, middleware, and communication channels.

Umair Khalid, et al. [14] decentralized lightweight blockchain-based authentication mechanism for IoT systems In IoT, heterogeneous technologies were used to create smart, ubiquitous items that were effortlessly connected to the Internet. Smart communities were just a few of the many areas in which these devices might be put to use. These Internet of Things (IoT) gadgets generate enormous amounts of data that must be protected for reasons of privacy and security. The security of these devices was therefore critical to the system's safety and performance.

Ahmad Al-Qerem, et al. [15] had changed to traditional concurrency control techniques in the cloud–fog systems can allow clients to avoid constantly communicating with the cloud server via the upstream communication channel. Through this study, a new version of the optimistic concurrency control protocol was introduced. It was possible to handle read-only IoT transactions at the fog node with the use of an augmented partial validation protocol. Only updates were forwarded to the cloud for final validation. Update transactions undergo partial validation at a fog node before being sent to a cloud, which makes them more opportunistic.

10.3. Problem Statement

Pattern recognition has been aided by algorithms such as SVM, CNN, and Random Forest, to name just a few [38, 39]. Most of the traditional research's goals have been achieved through current research. It was a review of the accuracy of the data classification. Each algorithm's efficiency and efficacy have been evaluated in terms of the accuracy of the data. CNN is better at analyzing and categorizing graphical data than SVM is at analyzing and categorizing textual data, based on our results from a literature review. There is hence a need for greater study on pattern detection models given the benefits of CNN. Despite these shortcomings, CNN's present system is capable of significant development. It takes a long time to compare graphical content in this manner. As a result, it is necessary to increase the performance of the standard CNN model. SVM was traditionally found to be the best method for text data analysis; however, convolution neural networks (CNN) have shown to be effective in picture analysis and classification. As a result, more work is needed on a pattern identification model to take advantage of CNN's advantages. SVM is not the ideal choice if the dataset is overlapping in some way. In some situations, Random Forest may outperform SVM in terms of performance.

10.4. Proposed Model

Image Processing is a method for enhancing raw photos collected from planes, spacecraft, and space probes, as well as images captured in everyday life. Over the last four to five decades, several methods for image processing have been developed.

There have been several advances in image processing for photos obtained by unmanned spacecraft, space probes, and military surveillance planes. It's becoming more and more popular for people to use image processing systems because of the rise of powerful computers, large memory devices, and other graphical tools. Cameras, such as those used to record video and take still images, are among the most common image acquisition systems.

Optical or analog camera images can be used as the input for the algorithm. Cameras with USB ports, such as CCD or CMOS sensors, are considered digital cameras. MATLAB presently uses augmented reality to record live video streams from the real environment. To access or communicate with this computer, MATLAB makes use of built-in tools called adapters that are attached directly to the PC (Image Processor). Removing the functional information from the captured images is one way to perform image processing. There are various robust features of an object, such as color and pattern, borders, strength, and form that must be noted if an object is to be characterized. The Intelligent Protection System makes use of a microprocessor designed for control purposes. A camera may be made to concentrate on just one picture at a time using this method. The camera consistently and constantly takes pictures at predetermined intervals.

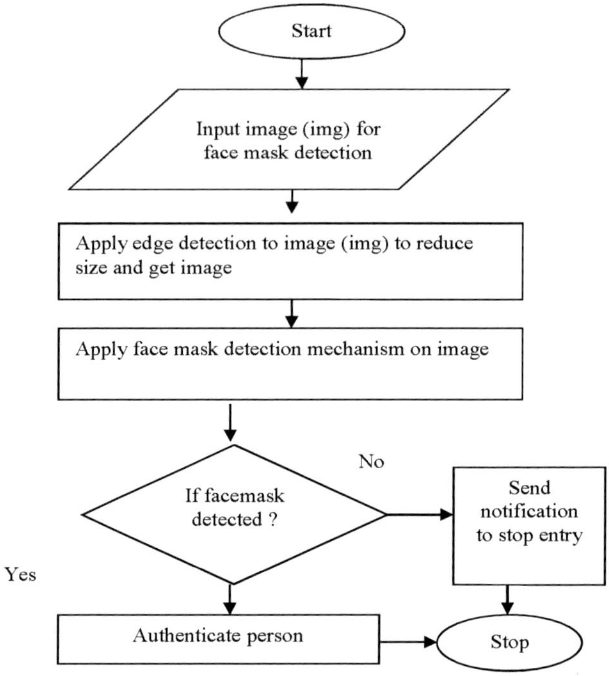

Figure 10.6. Flow chart of proposed work.

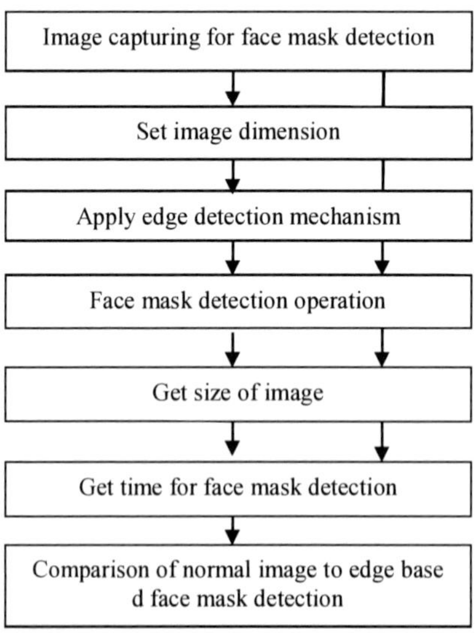

Figure 10.7. Proposed model.

10.4.1. Process Flow of Proposed Methodology

1. A camera-based data collection would be compiled into an image database. The image resizes function is used to preprocess camera-captured visual output.
2. After obtaining the picture dataset, run a traditional CNN classifier to see how much space and time it takes. In the event of an edge detection mechanism, the stored time and space variable may be compared to the incoming time and space variable.
3. Use the edge detection method to analyze the images. The edge detector would remove the unnecessary parts of the picture. The file size and time required for feature extraction are both calculated using edge detection.
4. Consider how much space and time the suggested CNN classifier would take up. There are supposed to be a lot of options in CNN's classified section. The picture collection may be adequately trained using the previously taught CNN classifier. A picture collection that has been trained is used to make judgments.
5. Compare the conventional and suggested work in terms of performance and space usage.

10.5. Result and Discussion

10.5.1. Implementation

The individual who isn't wearing a mask is seen in the next shot. Feature extraction from the colored picture necessitates a lengthy time for tracing. The individual wearing the mask is seen in the following picture. But the tracing takes a long time as the colored image takes a long time during feature extraction.

The edge detection method has been applied to the graphical image as seen in the accompanying picture. As the rendered graphical contents have been eliminated the image takes less space. More features are being extracted in a shorter amount of time than before. The feature extraction is fast as the edge-based image is taking relatively less time.

During CNN training image set of 15000 people has been considered where 7500 images were with a facemask and 7500 images were without a facemask. During training 2000 images were considered to check the accuracy of the model. The confusion matrix in the case of non-edge detected images is shown in Table 10.2 with the metrics shown in Table10.3.

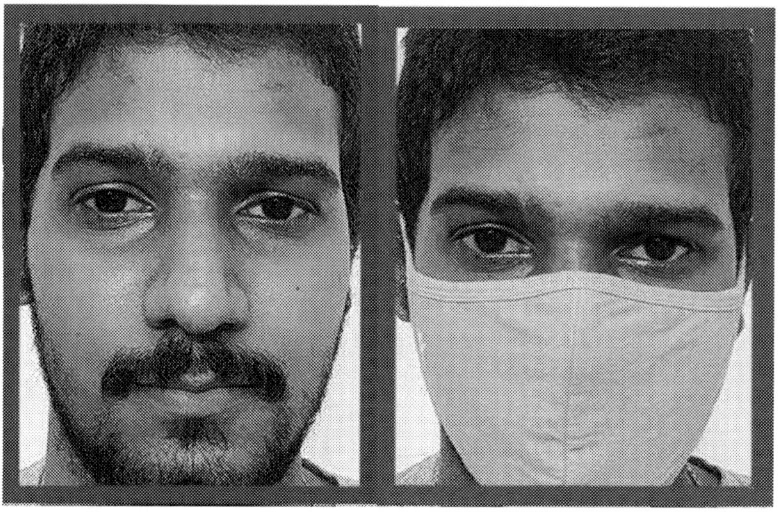

Figure 10.8. Before applying Edge Detector.

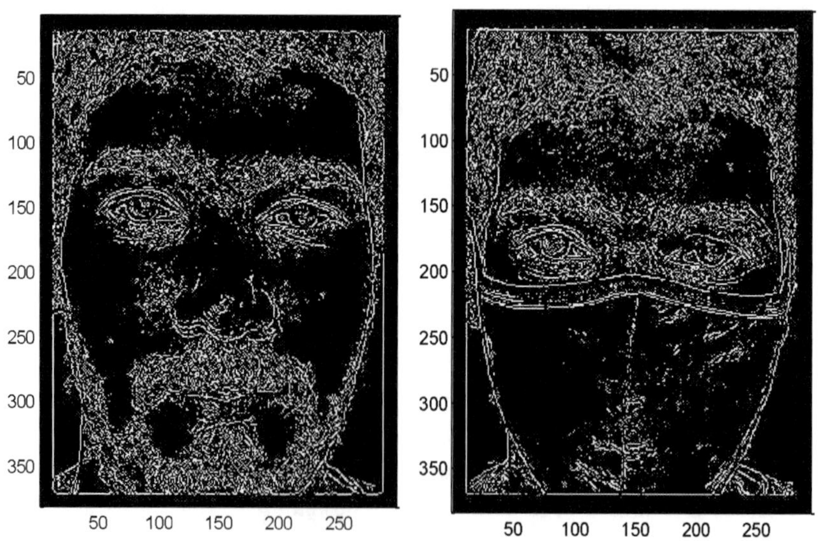

Figure 10.9. After applying Edge Detector.

Table 10.2. Confusion matrix in case of non-edge detected **images**

	Face Mask	**No face mask**
Facemask	1830	170
No Face mask	255	1745

Result
TP: 3575
Overall Accuracy: 89.38%

Table 10.3. Metrics for non-edge detected images

Class	**n(truth)**	**n(classified)**	**Accuracy**	**Precision**	**Recall**	**F-Score**
1	2085	2000	89.38%	0.92	0.88	0.90
2	1915	2000	89.38%	0.87	0.91	0.90

The confusion matrix in the case of the edge detected images are shown in Table 10.4 with its metrics shown in Table 10.5.

Table 10.4. Confusion matrix in case of the edge detected images

	Face Mask	**No face mask**
Facemask	1905	95
No Face mask	55	1945

Result
TP: 3850
Overall Accuracy: 96.25%

Table 10.5. Metrics for edge detected images

Class	**n(truth)**	**n(classified)**	**Accuracy**	**Precision**	**Recall**	**F-Score**
1	1960	2000	96.25%	0.95	0.97	0.96
2	2040	2000	96.25%	0.97	0.95	0.96

10.5.2. Comparative Analysis

1. Accuracy
2. Precision
3. Recall Value
4. F1-Score

Table 10.6. Comparative analysis of accuracy

Non Edge Detected	Edge Detected
89.38%	96.25%
89.38%	96.25%

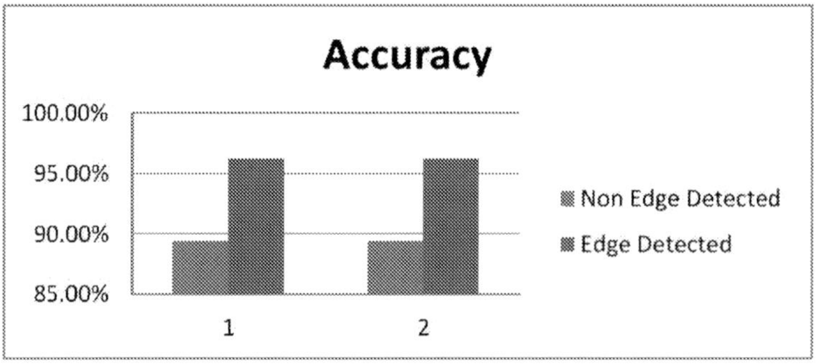

Figure 10.10. Comparative analysis of accuracy.

Table 10.7. Comparative analysis of precision

Non Edge Detected	Edge Detected
0.92	0.95
0.87	0.97

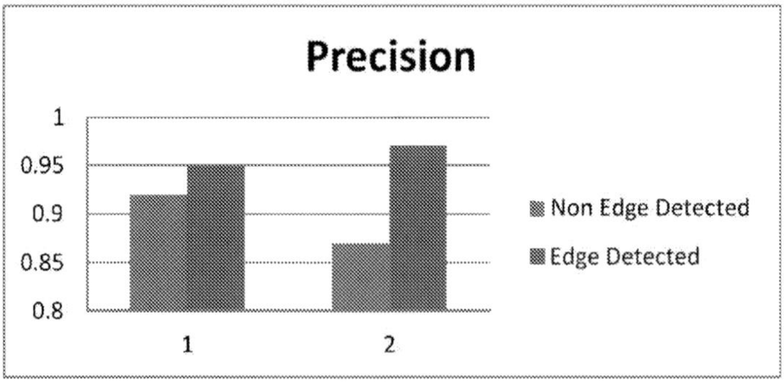

Figure 10.11. Comparative analysis of precision.

Table 10.8. Comparative analysis of Recall Value

Non Edge Detected	Edge Detected
0.88	0.97
0.91	0.95

Table 10.9. Comparative analysis of Fi-Score

Non Edge Detected	Edge Detected
0.90	0.96
0.90	0.96

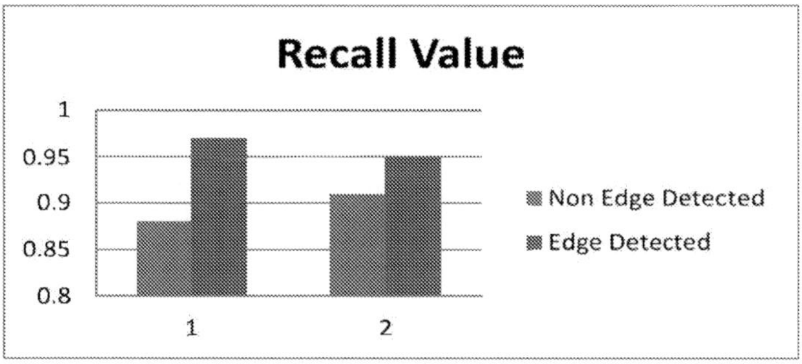

Figure 10.12. Comparative analysis of Recall Value.

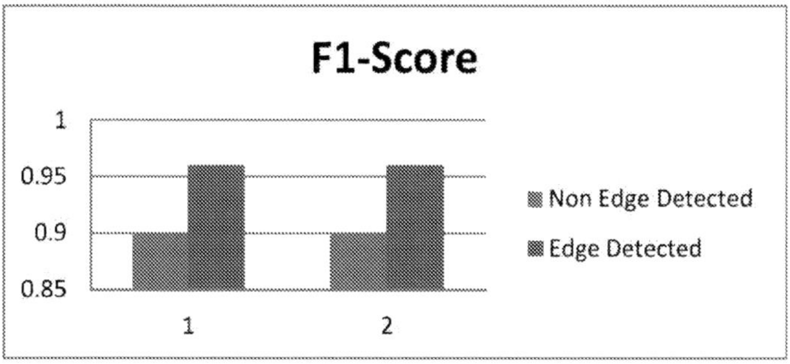

Figure 10.13. Comparative analysis of F1-Scope.

10.5.3. Simulation for Time Comparison

The edge-based CNN implementation is less time-consuming than the conventional CNN comparison. Comparing standard and suggested approaches, we can see how much time each would take.

Table 10.10. Simulations for time comparison

Frames	Time is taken in the previous model	Time is taken in the proposed model
100	185.2921315	88.27791
200	383.9363365	137.298
300	561.0483287	247.1158
400	709.4677632	255.4848
500	862.0706305	387.5247
600	1482.443749	727.2124
700	1134.156264	450.8819
800	1617.879617	737.4729
900	2049.912734	737.1776
1000	2375.860573	1177.339

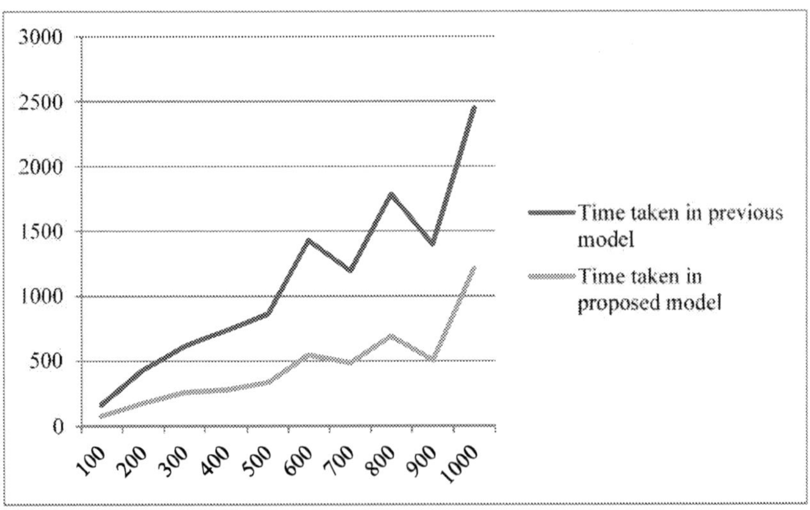

Figure 10.14. Simulation of time consumption in the case of traditional and proposed approach.

10.5.4. Simulation for Space Consumption

In addition, edge detection reduces the amount of storage space needed for data collection. Following is a graph comparing the amount of space used by both the suggested and conventional approaches.

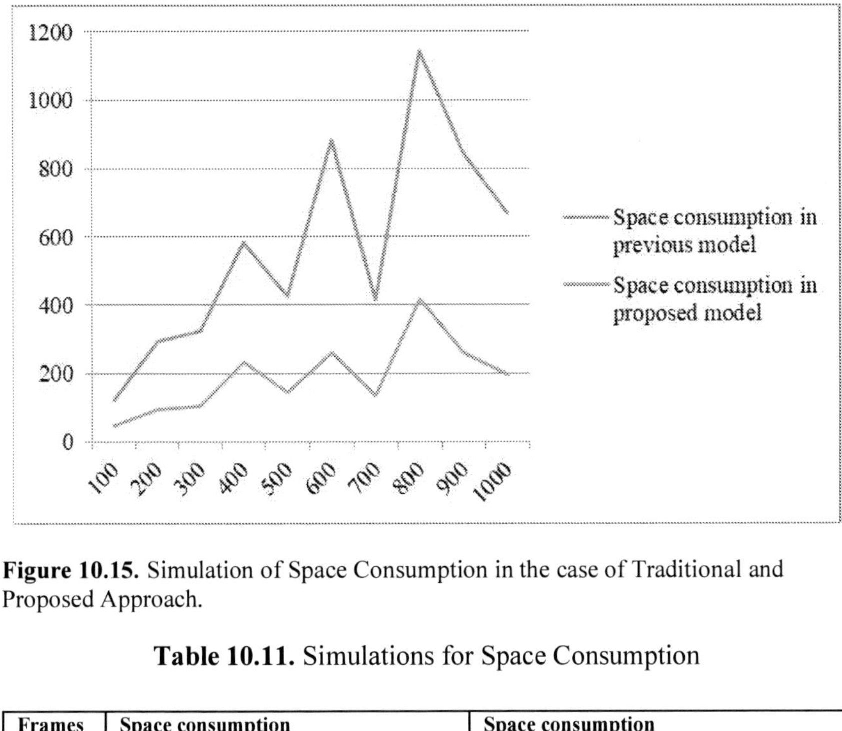

Figure 10.15. Simulation of Space Consumption in the case of Traditional and Proposed Approach.

Table 10.11. Simulations for Space Consumption

Frames	Space consumption in the previous model	Space consumption in the proposed model
100	120.4776613	46.99278045
200	291.6438095	93.67427571
300	323.4250898	104.4017499
400	583.4512701	232.5229961
500	426.3881185	146.0532324
600	883.7402449	259.5612359
700	414.5884674	132.9779225
800	1142.770745	419.5791016
900	847.666172	259.6798308
1000	668.3636602	193.9508764

10.5.5. Simulation of Accuracy

Before and after edge detection, the simulation of accuracy has been found here. To verify the correctness of the comparison, a sample of insitu was obtained.

10.5.6. Comparison of Proposed Work with Existing Technology

The comparison of the proposed work has been made to previous research considering time, space, accuracy, flexibility, scalability, and performance factors.

Table 10.12. Comparison chart for proposed and previous research

Feature	Previous Research	Proposed Research
Detection time	High	Low
Space	More	Less
Accuracy	less	high
Edge Detection mechanism	Not applied	Canny edge detection
Neural Network	Not applied	CNN
Flexibility	Less	High
Scalability	Limited	Huge scale
Performance	Low	High

Conclusion

Edge detection mechanisms have been shown to reduce the amount of time it takes to process data. The suggested model outperforms the prior CNN model in terms of overall performance. It also reduces the amount of RAM that's needed to run the app. Edge detection has increased CNN's capabilities since it reduces the size of graphical content and eliminates a lot of unnecessary graphical content. Simulation has a 14 percent higher accuracy rating than the existing system currently in use. As the size of the image increases, we see a shift in the precision of the measurement. In addition, changes in the dataset of the image have an impact. SVM has already been claimed to be appropriate for textual data by other researchers. However, when it comes to visual content, CNN comes out on top. Such efforts would be made to improve the performance of face mask detection. The advantages of CNN have been studied in depth. During face mask identification, CNN makes use of layers that can be utilized to examine characteristics of a data set that includes graphical content. The efficiency of the current CNN model has been improved by the usage of edge detection mechanisms. An edge detection algorithm is used during image segmentation to remove any extraneous graphic elements. In addition, research into graphical pattern detection models is sparse. As a result, researchers are focusing on edge-based CNN for graphical content processing.

Future Scope

An IoT-based project could benefit from the research's canny edge detection and CNN model. Such improvements are required for initiatives relying on image processing. In addition, the amount of space occupied by the graphical sample was manageable. +After making these changes, the classification and prediction system would have no effect. The suggested effort in medical research has the potential to improve CNN's decision-making capabilities. Accuracy may vary depending on image size; future work is expected to be more accurate. In addition, future studies will examine new graphical content adjustments. The scope based on the IoT provides new methods for organizations to produce value, although the continual connection and data exchange also present new potential for information to be compromised.

References

[1] Muthu, B. A., Sivaparthipan, C. B., Manogaran, Gunasekaran, Sundarasekar, Revathi, Kadry, Seifedine, Shanthini, A. & Dasel, Antony. "IOT based wearable sensor for diseases prediction and symptom analysis in the healthcare sector," *Peer-to-Peer Netw. Appl.,* vol. 13, no. 6, pp. 2123–2134, 2020, doi: 10.1007/s12083-019-00823-2.

[2] Atlam, H. F. and Wills, G. B. *IoT Security, Privacy, Safety, and Ethics.* Springer International Publishing, 2020.

[3] Pavithran, D., Shaalan, K., Al-Karaki, J. N. and Gawanmeh, A. "Towards building a blockchain framework for IoT," *Cluster Comput.*, vol. 23, no. 3, pp. 2089–2103, 2020, doi: 10.1007/s10586-020-03059-5.

[4] Al-Turjman, F. and Alturjman, S. "5G/IoT-enabled UAVsfor multimedia delivery in industry-oriented applications," *Multimed. Tools Appl.*, vol. 79, no. 13–14, pp. 8627–8648, 2020, doi: 10.1007/s11042-018-6288-7.

[5] Mabodi, K., Yusefi, M., Zandiyan, S., Irankhah, L. and Fotohi, R. "Multi-level trust-based intelligence schema for securing of internet of things (IoT) against security threats using cryptographic authentication," *J. Supercomput.*, vol. 76, no. 9, pp. 7081–7106, 2020, doi: 10.1007/s11227-019-03137-5.
[6] Kaur, M. J., Mishra, V. P. and Maheshwari, P. *The Convergence of Digital Twin, IoT, and Machine Learning: Transforming Data into action.* Springer International Publishing, 2020.
[7] Safara, F., Souri, A., Baker, T., Al Ridhawi, I. and Aloqaily, M. "PriNergy: a priority-based energy-efficient routing method for IoT systems," *J. Supercomput.*, vol. 76, no. 11, pp. 8609–8626, 2020, doi: 10.1007/s11227-020-03147-8.
[8] Islam, M. M., Rahaman, A., and Islam, M. R. "Development of Smart Healthcare Monitoring System in IoT Environment," *SN Comput. Sci.*, vol. 1, no. 3, 2020, doi: 10.1007/s42979-020-00195-y.
[9] Al Shorman, A., Faris, H. and Aljarah, I. "Unsupervised intelligent system based on one-class support vector machine and Grey Wolf optimization for IoT botnet detection," *J. Ambient Intell. Humaniz. Comput.*, vol. 11, no. 7, pp. 2809–2825, 2020, doi: 10.1007/s12652-019-01387-y.
[10] Verma, A. and Ranga, V. "Machine Learning-Based Intrusion Detection Systems for IoT Applications," *Wirel. Pers. Commun.*, vol. 111, no. 4, pp. 2287–2310, 2020, doi: 10.1007/s11277-019-06986-8.
[11] Vishwakarma, R. and Jain, A. K. "A survey of DDoS attacking techniques and defense mechanisms in the IoT network," *Telecommun. Syst.*, vol. 73, no. 1, pp. 3–25, 2020, doi: 10.1007/s11235-019-00599-z.
[12] Al-Emran, M., Malik, S. I. and Al-Kabi, M. N. A Survey of Internet of Things (IoT) in *Education: Opportunities and Challenges,* vol. 846. Springer International Publishing, 2020.
[13] Bansal, S. and Kumar, D. "IoT Ecosystem: A Survey on Devices, Gateways, Operating Systems, Middleware, and Communication," *Int. J. Wirel. Inf. Networks,* vol. 27, no. 3, pp. 340–364, 2020, doi: 10.1007/s10776-020-00483-7.
[14] Khalid, U., Asim, M., Baker, T., Hung, P. C. K., Tariq, M. A., and Rafferty, L. "A decentralized light weight blockchain-based authentication mechanism for IoT systems," *Cluster Comput.*, vol. 23, no. 3, pp. 2067–2087, 2020, doi: 10.1007/s10586-020-03058-6.
[15] Al-Qerem, A., Alauthman, M., Almomani, A. and Gupta, B. B. "IoT transaction processing through cooperative concurrency control on the fog–cloud computing environment," *Soft Comput.*, vol. 24, no. 8, pp. 5695–5711, 2020, doi: 10.1007/s00500-019-04220-y.
[16] Tseng, L., Yao, X., Otoum, S., Aloqaily, M. and Jararweh, Y. "Blockchain-based database in an IoT environment: challenges, opportunities, and analysis," *Cluster Comput.*, vol. 23, no. 3, pp. 2151–2165, 2020, doi: 10.1007/s10586-020-03138-7.
[17] Tun, S. Y. Y., Madanian, S. and Mirza, F. "Internet of Things (IoT) applications for elderly care: a reflective review," *AgingClin. Exp. Res.*, vol. 33, no. 4, pp. 855–867, 2021, doi: 10.1007/s40520-020-01545-9.

[18] Velnath, R., Prabhu, V. and Krishnakumar, S., 2021, March. Analysis of EEG Signal for the Estimation of Concentration Level of Humans. In *IOP Conference Series: Materials Science and Engineering* (Vol. 1084, No. 1, p. 012003). IOP Publishing.

[19] Schiliro, F., Beheshti, A., Ghodratmana, S., Amouzgar, F., Benatallah, B., Yang, J., Sheng, Q. Z., Casati, B. & Motahari-Nezhad, H. R. *ICOP: IoT-enabled policing processes,* vol. 11434 LNCS. Springer International Publishing, 2019.

[20] Ilapakurti, A. and Vuppalapati, C. "Building an IoT framework for connected dairy," *Proc. - 2015 IEEE 1st Int. Conf. Big Data Comput. Serv. Appl. BigDataService 2015*, pp. 275–285, 2015, doi: 10.1109/BigDataService.2015.39.

[21] Sansanwal, K., Shrivastava, G., Anand, R., & Sharma, K. (2019). Big data analysis and compression for indoor air quality. *Handbook of IoT and Big Data,* 1.

[22] Juneja, S., Juneja, A., & Anand, R. (2019, April). Reliability modeling for embedded system environment compared to available software reliability growth models. In*2019 International Conference on Automation, Computational and Technology Management (ICACTM)* (pp. 379-382). IEEE.

[23] Sindhwani, N., Maurya, V. P., Patel, A., Yadav, R. K., Krishna, S., & Anand, R. (2022). Implementation of Intelligent Plantation System Using Virtual IoT. In*the Internet of Things and Its Applications*(pp. 305-322). Springer, Cham.

[24] Singh, H., Ramya, D., Saravanakumar, R., Sateesh, N., Anand, R., Singh, S., & Neelakandan, S. (2022). Artificial intelligence-based quality of transmission predictive model for cognitive optical networks. *Optik*, *257*, 168789.

[25] Kohli, L., Saurabh, M., Bhatia, I., Sindhwani, N., &Vijh, M. (2021). Design and Development of Modular and Multifunctional UAV with Amphibious Landing, Processing and Surround Sense Module. *Unmanned Aerial Vehicles for Internet of Things (IoT) Concepts, Techniques, and Applications*, 207-230.

[26] Anand, R., Sinha, A., Bhardwaj, A., & Sreeraj, A. (2018). Flawed security of the social network of things. In *Handbook of research on network forensics and analysis techniques* (pp. 65-86). IGI Global.

[27] Yasumoto, K., Yamaguchi, H., & Shigeno, H. (2016). Survey of real-time processing technologies of IoT data streams. *Journal of Information Processing, 24*(2), 195-202.

[28] Bakshi, G., Shukla, R., Yadav, V., Dahiya, A., Anand, R., Sindhwani, N., & Singh, H. (2021). An Optimized Approach for Feature Extraction in Multi-Relational Statistical Learning. *Journal of Scientific and Industrial Research (JSIR)*, *80*(6), 537-542.

[29] Gupta, A., Srivastava, A., & Anand, R. (2019). Cost-effective Smart Home Automation using the Internet of Things. *Journal of Communication Engineering & Systems,* 9(2), 1-6.

[30] Shukla, R., Dubey, G., Malik, P., Sindhwani, N., Anand, R., Dahiya, A., & Yadav, V. (2021). Detecting crop health using machine learning techniques in a smart agriculture system. *Journal of Scientific and Industrial Research (JSIR)*, *80*(08), 699-706.

[31] Saini, M. K., Nagal, R., Tripathi, S., Sindhwani, N., & Rudra, A. (2008). PC Interfaced Wireless Robotic Moving Arm. In *AICTE Sponsored National Seminar on Emerging Trends in Software Engineering* (Vol. 50).

[32] Pandey, B. K., Mane, D., Nassa, V. K. K., Pandey, D., Dutta, S., Ventayen, R. J. M., ... & Rastogi, R. (2021). Secure text extraction from complex degraded images by applying steganography and deep learning. In *Multidisciplinary Approach to Modern Digital Steganography* (pp. 146-163). IGI Global.

[33] Liu, X., & Ansari, N. (2019). Toward green IoT: Energy solutions and key challenges. *IEEE Communications Magazine*, *57*(3), 104-110.

[34] Parthiban, K., Pandey, D., & Pandey, B. K. (2021). Impact of SARS-CoV-2 in online education, predicting and contrasting mental stress of young students: a machine learning approach. *Augmented Human Research*, *6*(1), 1-7.

[35] Meslie, Y., Enbeyle, W., Pandey, B. K., Pramanik, S., Pandey, D., Dadeech, P., ... & Saini, A. (2021). Machine intelligence-based trend analysis of COVID-19 for total daily confirmed cases in Asia and Africa. In *Methodologies and Applications of Computational Statistics for Machine Intelligence* (pp. 164-185). IGI Global.

[36] Babu, S. Z. D., Pandey, D., & Sheik, I. (2020). Acts of COVID19. *International Journal of Advanced Engineering*, *3*, 2457-0397.

[37] Pandey, D., Babu, S. Z. D., Noibi, T. O., Pandey, B. K., Dey, P. T. D., Kassaw, C., ... & Canete, J. J. O. (2020). A study on novel covid-19 epidemic threat for the universal physical situation in 2020. *Asian Journal of Advances in Research*, 27-33.

[38] Gupta, A., Anand, R., Pandey, D., Sindhwani, N., Wairya, S., Pandey, B. K., & Sharma, M. (2021). Prediction of Breast Cancer Using Extremely Randomized Clustering Forests (ERCF) Technique: Prediction of Breast Cancer. *International Journal of Distributed Systems and Technologies (IJDST)*, *12*(4), 1-15.

[39] Singh, H., Rehman, T. B., Gangadhar, C., Anand, R., Sindhwani, N., & Babu, M. (2021). Accuracy detection of coronary artery disease using machine learning algorithms. *Applied Nanoscience*, 1-7.

Chapter 11

A Study of Mental Stress of Online Learning during COVID-19

Rekha Narang[1,*], Prerna Dawar[1,†], Ritu Singh[1,#] and A. Shaji George[2,‡]

[1]Geeta University, Panipat, Haryana, India

[2]Department of Information and Communication Technology, Crown University, Int'l. Chartered Inc. (CUICI) Argentina Campus, Argentina

Abstract

The world is likely to experience a major health crisis in 21st century history. The corona virus (COVID-19) has put all nations on its knees and shut down various sectors and industries, including education. Teachers, students, and other education stakeholders are currently embroiled in controversy over what will happen to the education system if a majority of governments around the world are given the power to divide society as part of a 'new norm.' It is for this reason that educational institutions are developing an online platform as an effective platform for teaching reading and ensuring that students continue to pursue their studies of various social boundaries but this has impact on the mental stress of students. Our present study tries to understand how online learning has an impact on the mental state of students as according to their different demographic profiles. The sample of 150 students of different education level have been taken. The study found that there is no impact of gender, area, age, and education level except Education year

* Corresponding Author's Email: dr.narang12@gmail.com.

† Corresponding Author's Email: prernadawarsaluja@gmail.com.

Corresponding Author's Email: ritulohra1412@gmail.com.

‡ Corresponding Author's Email: drashajigeorge@gmail.com.

In: An Interdisciplinary Approach in the Post-COVID-19 Pandemic Era
Editors: D. Pandey, B. Kumar Pandey, N. Sindhwani et al.
ISBN: 979-8-88697-052-4

on mental stress of the students, ANOVA & t-test is used for the analysis through SPSS software.

Keywords: mental stress, COVID, online learning

Introduction

We all know that human beings are unique. Education is the important basis of this human life. Education has given a new direction to human life and provides a new thinking as a result of which human development takes place. When the light of education falls on the mirror of human form, the rays reflected from it remove the ignorance from the society. Knowing and understanding the importance of education is not the same, but this is the medium, with the help of which we can achieve the greatest achievements in our lives.

Real meaning of education is learning. To achieve the goal of our life, we should keep all those facts in our life. Education gives us the strength to face the biggest challenges of life with patience. Through education, we also get education to live in the society and understand our duties towards our family, our country at the personal level due to which we are able to perform our duty with full devotion and full dedication. In 2019, a new crisis took its toll, which was named as COVID Pandemic [1-4]. Due to this pandemic, there were so many changes in the field of education. No one would guess that the virus like COVID-19 would come and alter the lifestyle of people. Many difficulties were faced by the people. In COVID -19 session, government gave order to closer the school, colleges, educational institute, universities to prevent people from suffering this. With the closure of educational institutions, the birth of modernity in education took place. Students now started doing their studies with the help of digital classes [5, 6]. Due to the promotion of modernity in the field of education, new experiments were also being done in it. Teachers should give any information related to the subject to their students. We came to know about its positive effects when modernity was created in the field of education. On the other hand, its negative effects were also born. This had a negative effect on the mind of the children such as mental stress, long screen time, lack of confidence, difficulty in maintaining concentration.

In fact, the beneficiaries were left uncertain about their studies and stressed how they could cope with the new trends introduced by the epidemic. However, in order to alleviate the anxiety that students experience in many educational institutions consider, some changes are done in the entire education system so that learning is not compromised during the global decision to be isolated from society. The main goal of this transformation of the education system is to facilitate learning in the comfort of student homes. Therefore, almost all educational institutions develop home education and develop modules, methods, and methods of how this program can work to ensure that students are still able to meet the basic educational needs imposed on their schools, colleges, and universities. Synchronized learning is very important for all learning. However, in this epidemic, it is better to have new ways and means to create and facilitate learning so that there is no one left and no better way than to remove learning closure and comfort of one's home, making it an educational institution. However, in order to further facilitate learning and meet the educational needs of students while at home, educational institutions have adopted eLearning or online classes.

E-Learning is "a form of learning designed to create an online connection between teacher and student" that makes the home an extension of the learning environment. In addition, some people argued that in making the home a separate educational institution, parents and adults have a responsibility to help students make a better learning environment [7, 8]. At e-Learning, parents and adults become learning facilitators in the absence of the teacher. Typically, e-learning aims to complement the learning or learning process [9, 10] that takes place within the classroom so that the learning process does not stop within the classroom but continues into the home environment. This makes learning more powerful.

During the pre-epidemic situation, e-learning is only close and second to distance learning; it is not intended to replace the distance learning. However, in the event of an ongoing global catastrophe due to COVID-19 disease, e-learning may stop compliance with distance learning; it may be, in a new normal way. Practical methods are not the usual distance learning methods to ensure that students' education is not disrupted and that their educational needs are met. This has forced students to switch from online learning to formal and informal classroom teaching. This change in the way, the education sector is done, is one of the many ways to help students whose academic work is affected by the epidemic. However, this change from online learning is not free from problems. While it has a positive impact on the learning system especially during this epidemic, students still face difficulties and pressures in this new way of learning. Online learning has a positive and negative side to how students go through the process. It is clear that this study wishes to expose itself. Therefore, this study focuses on the stress caused by the epidemic among students as they continue their academic career using online learning as a tool to pursue their learning efforts in view of the social limitations of this new standard.

Review of Literature

A vision of academic oppression and strategies for dealing with emotional intelligence among college students [11] has led to research current and coping strategies. The findings of this paper will be very much useful in exploring important online education challenges especially in the era of COVID-19. The author described in the study that students feel highly depressed in Online education during COVID-19.

The study in [12] aimed at comparing mood, suffocated responses, sleep patterns, and behavior of people with non-disease provinces in China. Within two weeks, senior health workers and People in Hubei Province were less concerned on mental health outcomes among primary and secondary health workers during the coronavirus 2019 violence COVID-pandemic in Italy.

The work in [13] studied to identify the psychological impact of student learning on the internet during the COVID-19 epidemic and found that the burden of online speech activities is one of the factors that contributes to students' stress levels, which they need to use.

One more study done by the research in [14] on mental health and online learning during the COVID-19 epidemic found that the developing world is needed to reduce the stress that comes with online learning, including video-conferencing. In the case of COVID-19, we need existing social considerations inequalities and economic barriers to protecting their mental health from online learning.

The paper [15] highlights the effect of online learning on the mental health of tertiary students and effect of Online Learning for Higher Education Students during COVID-19 Lockdown. Some great findings from this research have shown that stress in the students is due to the online learning even though it is a solution to the challenges of shutting down COVID-19 learning.

The objectives of the study in [16] have considered all the steps taken by the Government of India to support the online learning process during the closure of COVID-19. It was stated that online earning can play an important role in keeping people busy and free from stress.

In [17], it has been studied the psychological effects to underline the introduction of online learning during COVID-19 Visual stress among students in visual classes during the COVID-19 outbreak in KSA. The barriers to online learning have also been highlighted during the COVID-19 epidemic. It stated that students feel highly pressurized in educational institutions in the Republic of Kazakhstan switch to online learning (OL).

The purpose of another study [18] was to investigate mental stress, and the fear of losing a year of study. It was an important factor contributing to depression during the closure of pandemic described by Online learning. We must be thankful to COVID-19 improved mental health among medical students.

In [19], the study of Online Learning on attitudes of students in Jordanian University is considered. Jordanian University Student have given opinions on Emergency Learning in online mode during COVID-19. According to this study, psychological stress has reduced students' willingness to study online in the future as it becomes the need to transit on online learning.

In [20], the psychological impact of digital transformation is studied as a result of the COVID-19 epidemic, that has led to rapid online learning at undergraduate level. COVID-19 poses a legal threat to mental health throughout the nation from its inception. The study aimed to explore the underlying causes of online learning stress, and finally, students' attitudes toward online learning across the pandemic.

The authors in [21] have done a joint study on the various perceptions like stress, anxiety and mental health in online learning in colleges and schools in India during COVID-19. This survey included one suggested question where parents can offer their suggestions for dealing with online learning during COVID-19. Descriptive statistics I has been used such as frequency, Mean, standard deviation and percentage for estimation.

In [22], the negative psychological impact of this situation on the students, such as having the right internet during the first week of internet launch study at KSU is considered. Further research is needed to understand what students are going through in various fields and across in Online learning during the COVID-19 pandemic and the educational pressure of university students.

The authors in [23] proposed policies and procedures in many areas to make education more inclusive and accessible to all readers. Enhancing the professional capacity of ordinary and special school teachers in delivering online teaching is focused with special focus on students from the undergraduate and CWD categories. Improving access to electronic media and internet communication in all parts of the country and in various socio-economic sectors and Education during the COVID-19 epidemic and its psychological impact on children and youth in India has also been focussed.

In response to the COVID-19 epidemic, more than a thousand schools, colleges and universities all over the world closed their doors in March 2020. Millions of students have completed their semester of high school, which led to an unusual disruption of education in the United States. Although COVID-19 poses a small risk to the health and death of senior college students, the pandemic is likely to cause significant uncertainty and depression [24].

The authors in [25] studied the impact of demographic variables on the perceived stress amongst professionals from different industries and found that demographic variables like age, gender, marital status and current location have no impact on stress but working hours have a significant impact on stress.

Objectives of the Study

There are so many objectives of the present study:

- To find the online learning impact on the mental stress level of students during COVID-19.
- To find the online learning impact on the mental stress according to gender and age of students.
- To find the online learning impact on the mental stress according to residential areas of students.
- To find the online learning impact on the mental stress according to education year and education level of students.

Research Methodlogy

The population chosen for this study has been from the students of different education level. The sample size was 150. The scale used for the study was prepared on the basis of literature review. For the context and purpose of research in order to fit, appropriate modification were made. All the statements were measured on 5 points likert scale, ranging from 1 to 5 in strongly disagree to strongly agree scale.

- *Study Area*: The study was conducted in senior secondary students, graduate students and post graduate students of Panipat, Haryana, India.

- Research hypothesis:
  - H01 Gender has no significant difference on mental stress in online learning.
  - H02 Age has no significant difference on mental stress in online learning.
  - H03 Area has no significant difference on mental stress in online learning.
  - H04 Education has no significant difference on mental stress in online learning
  - H05 education year has no significant difference on mental stress in online learning.

Table 1 and Table 2 indicate that there is no statistically significant difference in mental stress level of students based on the gender, t(97) = -1.291, p = 0.200. Hence it is observed that Mental Stress level of both male and female students are same.

Table 3 and Table 4 show that there is no significant impact of Education level of students on Mental Stress of students due to online learning, F (2,96) = 0.014, p = 0.986. Hence it is observed that education level of students does not affect the stress perceived by students.

Table 5 and Table 6 show that there is a significant impact of Education year on Mental Stress of Students, F (3,95) = 2.591, p = 0.047 Hence it is observed that education year of the students affect their mental status and stress perceived by students of different education year is not same.

Table 7 shows the independent samples test. Table 8 and Table 9 show that there is no significant impact of Age on Mental Stress of students due to online learning, F (3,95) = 0.516, p = 0.673. Hence it is observed that age of the students does not affect their mental stress level and stress perceived by students of different age group.

Table 10 and Table 11 indicate that there is no statistically significant difference in mental stress level of students based on the area, t (97) =-0.559, p = 0.578. Hence it is observed that Mental Stress level of any area of students is same.

Table 1. Group statistics

	Gender	N	Mean	Std. Deviation	Std. Error Mean
Mean	Male	59	3.5435	0.63565	0.08275
	Female	40	3.7117	0.63658	0.10065

Table 2. Independent Samples Test

	Levene's Test for Equality of Variances		t-test for Equality of Means						
Mean	**F**	**Sig.**	**T**	**df**	**Sig. (2-tailed)**	**Mean Difference**	**Std. Error Difference**	**95% Confidence Interval of the Difference**	
								Lower	**Upper**
Equal variances assumed	0.658	0.419	-1.291	97	0.200	-0.16822	0.13027	-0.42676	0.09033
Equal variances not assumed			-1.291	83.799	0.200	-0.16822	0.13030	-0.42735	0.09091

Table 3. Descriptives

Mean	N	Mean	Std. Deviation	Std. Error	95% Confidence Interval for Mean		Minimum	Maximum
					Lower Bound	Upper Bound		
Senior Secondary	19	3.5951	.82209	.18860	3.1989	3.9914	2.00	5.00
Under Graduate	40	3.6234	.57977	.09167	3.4380	3.8088	2.24	4.71
Post Graduate	40	3.6072	.61243	.09683	3.4113	3.8030	1.94	4.88
Total	99	3.6114	.63818	.06414	3.4841	3.7387	1.94	5.00

Table 4. ANOVA

Mean	Sum of Squares	Df	Mean Square	F	Sig.
Between Groups	0.012	2	0.006	0.014	0.986
Within Groups	39.902	96	0.416		
Total	39.913	98			

Table 5. Descriptives (Graduate Level students)

Mean	N	Mean	Std. Deviation	Std. Error	95% Confidence Interval for Mean		Minimum	Maximum
					Lower Bound	Upper Bound		
Ist Year	74	3.5302	0.64196	0.07463	3.3815	3.6790	1.94	4.88
IInd Year	11	3.6257	0.46113	0.13904	3.3159	3.9355	3.12	4.53
IIIrd Year	11	4.0695	0.67163	0.20250	3.6183	4.5207	3.00	5.00
IV year	3	3.8824	0.29412	0.16981	3.1517	4.6130	3.59	4.18
Total	99	3.6114	0.63818	0.06414	3.4841	3.7387	1.94	5.00

Table 6. ANOVA (Graduate Level Students)

Mean	Sum of Squares	df	Mean Square	F	Sig.
Between Groups	3.019	3	1.006	2.591	0.047
Within Groups	36.895	95	0.388		
Total	39.913	98			

Table 7. Independent samples test

		Levene's Test for Equality of Variances		t-test for Equality of Means						
		F	Sig.	t	df	Sig. (2-tailed)	Mean Difference	Std. Error Difference	95% Confidence Interval of the Difference	
									Lower	Upper
Mean	Equal variances assumed	1.360	0.246	0.559	97	0.578	0.07209	0.12906	-0.18405	0.32823
	Equal variances not assumed			0.564	96.997	0.574	0.07209	0.12780	-0.18156	0.32574

Table 8. Descriptives

Mean	N	Mean	Std. Deviation	Std. Error	95% Confidence Interval for Mean		Minimum	Maximum
					Lower Bound	Upper Bound		
Below 15	3	4.0196	0.14804	0.08547	3.6519	4.3873	3.88	4.18
15-20	49	3.5668	0.66640	0.09520	3.3754	3.7582	2.00	5.00
20-25	45	3.6273	0.63730	0.09500	3.4358	3.8188	1.94	4.88
Above 25	2	3.7353	0.20797	0.14706	1.8667	5.6039	3.59	3.88
Total	99	3.6114	0.63818	0.06414	3.4841	3.7387	1.94	5.00

Table 9. ANOVA

Mean	Sum of Squares	Df	Mean Square	F	Sig.
Between Groups	0.639	3	0.213	0.516	0.673
Within Groups	39.274	95	0.413		
Total	39.913	98			

Table 10. Group Statistics

	Residence	N	Mean	Std. Deviation	Std. Error Mean
Mean	Rural	53	3.6449	.67941	.09332
	Urban	46	3.5728	.59221	.08732

Table 11. Independent Samples Test

	Levene's Test for Equality of Variances		t-test for Equality of Means						
Mean	F	Sig.	t	df	Sig. (2-tailed)	Mean Difference	Std. Error Difference	95% Confidence Interval of the Difference	
								Lower	Upper
Equal variances assumed	1.360	.246	.559	97	.578	.07209	.12906	-.18405	.32823
Equal variances not assumed			.564	96.997	.574	.07209	.12780	-.18156	.32574

Limitation of the Study

The major limitation of the study is that sample size was very less as we considered here one stakeholder of online learning, that is only students. The future scope of the study will be to take other stakeholders also.

Conclusion

The present study is aimed to identify the impact of online learning on mental stress of students during COVID-19. Based on the analysis, it is concluded that there is no significance impact of gender, area, age and education level (except education year) on mental stress of the students. Final year students feel highly pressure during pandemic because of their last year of programme. They are concerned for their placements, career and future.

References

[1] Anand, R., Sindhwani, N., & Saini, A. (2021). Emerging Technologies for COVID-19. Enabling Healthcare 4.0 for Pandemics: A Roadmap Using AI, Machine Learning, *IoT and Cognitive Technologies*, 163-188.

[2] Meivel, S., Sindhwani, N., Anand, R., Pandey, D., Alnuaim, A. A., Altheneyan, A. S., & Lelisho, M. E. (2022). Mask Detection and Social Distance Identification Using Internet of Things and Faster R-CNN Algorithm. *Computational Intelligence and Neuroscience*, 2022.

[3] Pandey, D., Ogunmola, G. A., Enbeyle, W., Abdullahi, M., Pandey, B. K., & Pramanik, S. (2021). COVID-19: A framework for effective delivering of online classes during lockdown. *Human Arenas*, 1-15.

[4] Yadeta, A. E., & Pandey, D. (2020). Analysis of the global impact of the pandemic (COVID-19) on construction industry: possible scenarios. *Current Trends in Civil & Structural Engineering*, 6(4), 1-8.

[5] Pandey, D., Babu, S. Z. D., Noibi, T. O., Pandey, B. K., Dey, P. T. D., Kassaw, C., & Canete, J. J. O. (2020). A study on novel covid-19 epidemic threat for universal physical situation in 2020. *Asian Journal of Advances in Research*, 27-33.

[6] Rather, R. A., Islam, T., Rehman, I. U., & Pandey, D. (2021). Development of vaccine against coronavirus disease 2019 (COVID-19) In India. *Asian Journal of Advances in Medical Science*, 13-21.

[7] Jain, S., Sindhwani, N., Anand, R., & Kannan, R. (2022). COVID Detection Using Chest X-Ray and Transfer Learning. In *International Conference on Intelligent Systems Design and Applications* (pp. 933-943). Springer, Cham.

[8] Khongsai, L., Anal, T. S., AS, R., Shah, M., & Pandey, D. (2021). Combating the spread of COVID-19 through community participation. *Global Social Welfare*, 8(2), 127-132.

[9] Shukla, R., Dubey, G., Malik, P., Sindhwani, N., Anand, R., Dahiya, A., & Yadav, V. (2021). Detecting crop health using machine learning techniques in smart agriculture system. *Journal of Scientific and Industrial Research (JSIR),* 80(08), 699-706

[10] Singh, H., Rehman, T. B., Gangadhar, C., Anand, R., Sindhwani, N., & Babu, M. (2021). Accuracy detection of coronary artery disease using machine learning algorithms. *Applied Nanoscience*, 1-7.

[11] Chandra, Y. (2020). Online education during COVID-19: perception of academic stress and emotional intelligence coping strategies among college students. *Asian Education and Development Studies*.

[12] Yang, X., Xiong, Z., Li, Z., Li, X., Xiang, W., Yuan, Y., & Li, Z. (2020). Perceived psychological stress and associated factors in the early stages of the coronavirus disease 2019 (COVID-19) epidemic: Evidence from the general Chinese population. *PloS one*, 15(12), e0243605.

[13] Chinna, K., Sundarasen, S., Khoshaim, H. B., Kamaludin, K., Nurunnabi, M., Baloch, G. M., & Memon, Z. (2021). Psychological impact of COVID-19 and lock down measures: An online cross-sectional multicounty study on Asian university students. *PloS one*, 16(8), e0253059.

[14] Alibudbud, R. (2021). On online learning and mental health during the COVID-19 pandemic: Perspectives from the Philippines. *Asian Journal of Psychiatry*.

[15] Akpınar, E. (2021). The Effect of Online Learning on Tertiary Level Students Mental Health during the COVID-19 Lockdown. *The European Journal of Social & Behavioural Sciences*.

[16] Jena, P. K. (2020). Impact of pandemic COVID-19 on education in India. *International Journal of Current Research* (IJCR), 12.

[17] AlAteeq, D. A., Aljhani, S., & AlEesa, D. (2020). Perceived stress among students in virtual classrooms during the COVID-19 outbreak in KSA. *Journal of Taibah University Medical Sciences*, 15(5), 398-403.

[18] Perissotto, T., Silva, T. C. R. P. D., Miskulin, F. P. C., Pereira, M. B., Neves, B. A., Almeida, B. C., & Nunes, P. V. (2021). Mental health in medical students during COVID-19 quarantine: a comprehensive analysis across year-classes. *Clinics*, 76.

[19] Al-Salman, S., & Haider, A. S. (2021). Jordanian University Students' Views on Emergency Online Learning during COVID-19. *Online Learning*, 25(1), 286-302.

[20] Saha, A., Dutta, A., & Sifat, R. I. (2021). The mental impact of digital divide due to COVID-19 pandemic induced emergency online learning at undergraduate level: Evidence from undergraduate students from Dhaka City. *Journal of Affective Disorders*, 294, 170-179.

[21] Harjule, P., Rahman, A., & Agarwal, B. (2021). A cross-sectional study of anxiety, stress, perception and mental health towards online learning of school children in India during COVID-19. *Journal of Interdisciplinary Mathematics*, 24(2), 411-424.

[22] Moawad, R. A. (2020). Online learning during the COVID-19 pandemic and academic stress in university students. *Revista Românească pentru Educaţie*

Multidimensională [Romanian Journal for Multidimensional Education], 12(1 Sup2), 100-107.

[23] Mahapatra, A., & Sharma, P. (2021). Education in times of COVID-19 pandemic: Academic stress and its psychosocial impact on children and adolescents in India. *International Journal of Social Psychiatry*, 67(4), 397-399.

[24] Clabaugh, A., Duque, J. F., & Fields, L. J. (2021). Academic stress and emotional well-being in united states college students following onset of the COVID-19 pandemic. *Frontiers in Psychology*, 12.

[25] Gharia, R., Narang, R., & Karnik, M. P. (2020). Impact Of COVID 19 Lock Down On The Employees'well-Being: A Study With Reference To Professionals. *Palarch's Journal of Archaeology of Egypt/Egyptology*, 17(9), 2458-2472.

Chapter 12

The Impact of Digital Change on Student Learning and Mental Anguish in the COVID Era

Binay Kumar Pandey[1,*], Digvijay Pandey[2], Rohit Anand[3], Harinder Singh[4], Nidhi Sindhwani[5] and Yazusha Sharma[6]

[1]Department of Information Technology, College of Technology, Govind Ballabh Pant University of Agriculture and Technology, Pantnagar, India
[2]Department of Technical Education, IET, Dr. APJ Abdul Kalam Technical University, Lucknow, India
[3]Department of ECE, G. B. Pant DSEU Okhla-1 Campus (formerly GBPEC), New Delhi, India
[4]Department of CS and IT, Sant Baba Attar Singh Khalsa College, Sandaur, Punjab, India
[5]AIIT, Amity University, Noida, UP, India
[6]Jaipur Engineering College and Research Centre (JECRC), Jaipur, Rajasthan, India

Abstract

There is a lot of change in the learning of the students after the pandemic COVID-19. To study the resulting impact on their learning is the main aim of this article. To review this, a dataset of the various students is created and subsequently processed and visualized. The data is undergone to the various classification techniques using machine learning. It is observed after the analysis that the support vector machine (SVM) method is best in terms of the classification accuracy while random forest (RF) method is best in terms of the classification sensitivity.

* Corresponding Author's Email: digit11011989@gmail.com.

In: An Interdisciplinary Approach in the Post-COVID-19 Pandemic Era
Editors: D. Pandey, B. Kumar Pandey, N. Sindhwani et al.
ISBN: 979-8-88697-052-4

Keywords: COVID-19, student learning, classification, dataset, machine learning

1. Introduction

The coronavirus disease 2019 (COVID-19) [1] was firstly discovered in China in December 2019, quickly spread over the world, and was declared a pandemic by the World Health Organization on March 11, 2020. In the spring of 2020, many universities all around the world were required to close their campuses and move all of their academic programmes online [2]. Universities were unprepared for such a shift from traditional classroom instruction to entirely online instruction. Initially, most of the universities lacked facilities and plans [3].

In the recent few decades, there have been so many advancements in the various technologies [4-8], which have proven to be extremely valuable during this pandemic. There were several internet venues that assist online education [9]. Universities, on the other hand, found it difficult to map their educational activities in an online environment. Professors and students also confronted a slew of logistical, technical, financial, and social issues [10].

The pandemic and the ensuing lockdowns have had an impact on people's mental health all over the world. Many students suffer from worry and stress [11]. Students' ability to adjust to online education is frequently hampered by psychological concerns. Furthermore, not all students have the same level of access to and knowledge of digital technologies. Although these disparities existed before the COVID-19 pandemic, the COVID-19 pandemic has brought this digital gap to light [12]. COVID-19 has had surprising and major repercussions all across the world. It examines the immediate consequences of school closures and reopenings, as well as the long-term implications on students, their families, and the educational system as a whole.

The contribution of this article is to discuss the various after effects of the change in the education and digital technologies on the learning and mental behavior of the students as well as learners.

The chapter is organized as follows. Next section discusses the review of the literature related to the present work. The subsequent section discusses the methodology with the main emphasis on the various classification techniques. At last, a brief conclusion is drawn.

2. Related Work

The authors in [13] mentioned about the research done for bachelor and master students in the field of technology and engineering. Currently the world has entered into the digital one due to the issues caused by the pandemic COVID-19. Learning for the persons has been transformed entirely into e-learning that has impacted the educational institutes and organizations. This has resulted into the mixed teaching (having the combination of online and offline teaching) for the upcoming post-COVID era. The authors have done the data analysis in concerned with the e-sessions and offline sessions. It has been shown that the learners are agreed to the slightly negative impact of e-classes viz. lack of proper communication through online mode. In [14], a unique model with the complete framework and execution of e-classes having atmosphere similar to the offline classes has been presented. Both the administration and technical aspects have been taken care while designing this novel model. Further, one more paradigm has been framed to conduct the offline classes with proper distancing in post-COVID era. All the aspects of teaching and learning have been affected because of switching from offline classes to online classes as the online classes can't replicate the offline classes due to the huge difference in the quality of education in e-classes and offline classes. The authors in [15] have discussed the changes expected in the higher education system in the post-COVID era. Further, they have suggested a hybrid paradigm with the help of high level language C. A hybrid model has been suggested where 1st model refers to participatory and assessment model while 2nd model refers to an online course. The first model results in a logically sound classroom. With this new model, the learning can be optimized that stretches the teaching level as the classrooms are converted to multi-dimensional classrooms. A combined evaluation system focusing on formative evaluation and sum can be done. The analysis has shown that around 95% learners may be adapted to this hybrid model.

Further in [16], the authors discussed the switching from offline teaching to online teaching with collective techniques. It has been shown that the facilitator collected the various observations and the learners responded with the answers to the questions on the collective work after the complete session. The authors had shown the positive aspects of the collective online learning and discussed the issues arising while adapting to the atmosphere of teaching-learning system. The authors in [17] discussed the framework of distance learning in the COVID era. The major issue in this framework is the compromise between accomplishing the desired results and the occurring resources. Various resources like laptops, microcontrollers and mobile phones have been considered while implementing the various kinds of learning for complementing the class-based learning. These types of learning may be utilized to enhance the performances of students especially during the implementation of the project works. The observation of the authors in [18] refers to the primary issues to create student learning, to organize the various e-classes and to make the techniques for the best assessment. Various aspects were considered as per the various resources and learning aims. The analysis discussed in this research has been of major use in the learning while studying in the higher education institute.

In [19], the authors made a review regarding the contentment for the e-learning during COVID-19 period. The assessment and examination marks of the learners are found to be better in offline classes than e-classes. So, this work is focused towards enhancing the relationship between teachers and learners so as to inspire the learners and enhance their learning level with the help of big data technology. In brief, this work provides a basic way for the better evolution of e-teaching. The paper [20] discussed the reconnoitring of the applications of virtual reality (VR) to enhance the student learning while also discussing the issues in the application of the various virtual reality challenges. The collective learning during COVID-19 is very promising to assist the various students in developing their skills. In [21], the concept of artificial intelligence (AI) is suggested to conserve the outcomes of learning while following a set of rules during the pandemic. The concept of AI has been categorized into predictive, perspective and descriptive inspection. This study involves a collection of the various things learnt while using the various AI methods as well as tools. In [22], the concept of artificial intelligence and 5G is discussed in the enhancement of online learning. In this research, there has been the analysis of the various questionnaire reviews for influencing the online education. This research also finds out that the number of learners spending time on e-classes is going on increasing and hence some implications have been suggested.

3. Methodology

As of March, an educational place has been in upheaval as a result of various like the impending closure of schools and colleges owing to a disease epidemic [23-27]. As the school year draws to a close, there is legitimate worry about finishing the course. As a result, online classes such as Zoom, Youtube

Classroom, Webex Meet, Microsoft Meet, Teams, Meeting, and others have sprung up. Every coin has two sides: one side is a great chance to implement lessons online, while the other side is a range of challenges with students attending classes [28]. This is also a source of concern.

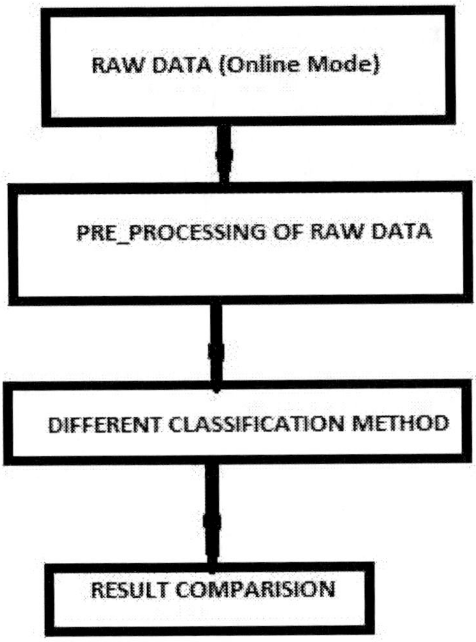

Figure 1. General machine learning methodology.

The categorization model is a type of data mining technique that breaks down provided data into separate instances. This allocates a certain component to an instance with a very low failure rate. It would definitely eliminate the need to represent critical information classifications inside such a given data collection. A classification algorithm is utilized to predict people's anxiety levels in this study [29, 30]. Training and testing are done with a 1:3 data ratio. A general methodology using Machine learning is shown in figure 1.

3.1. Data Collection and Pre-Process

The data was obtained from the responses of 649 understudies to online course ideas. A questionnaire was used to conduct the survey. Participants, or learners, are asked to complete a questionnaire based on their personal experiences. Participants' replies were divided into two categories: those gathered prior to online courses and those collected during online courses, with a total of 46 questions posed.

3.2. Data Preparation and Data Visualization

Following the data collection, a cleansing operation will begin to remove errors, duplicate entries, and inconsistent data. After that, visualising is a main goal. This is necessary in order to discover relationships between characteristics and conduct data exploration. Furthermore, data is separated into two categories: training data and testing data.

Table 1. Comparison of classification algorithms

Algorithm	Accuracy (%)	Sensitivity (%)
Random Forest	82.7	100
Support Vector Machine	86.85	75
Naive Bayes	73	66

3.3. Comparison of Different Classification

Classification is a distinguishing process across machine learning methodologies, as seen in Table 1. Students' comments were classified using methods such as Random Forests, k-nearest neighbours (k = 5), support vector machine and, naive Bayes, which were based on responses supplied largely during the COVID-19 shutdown's web-based debate [31, 32].

(a) *Random Forest.* It's true that the Random Forest algorithm is a tree-based ensemble. These are made up of a number of decisions that come together to form a forest. The Random Forest model may be used for both classification and regression. The algorithm as a whole handles a lot of data. To boost classification efficiency, we employed 600 decision trees and Ginni for imperfection indices in our research.

(b) *Support Vector Machine.* Image classification, text categorization, medical research, and physical machine diagnostic testing are just a few of the fields where SVM has been successfully and often used for empirically proven challenges. The usual response of this classifier is primarily hyper plane.

(c) *Naive Bayes.* Collaboration has referred to Naive Bayes classification algorithms as "classification algorithms." They function on the basis of probability. Naive Bayes classifiers have a wide range of applications. They are frequently projected to have been temporarily independent for just a single goal value.

Table 1 shows the comparison of different classification algorithms.

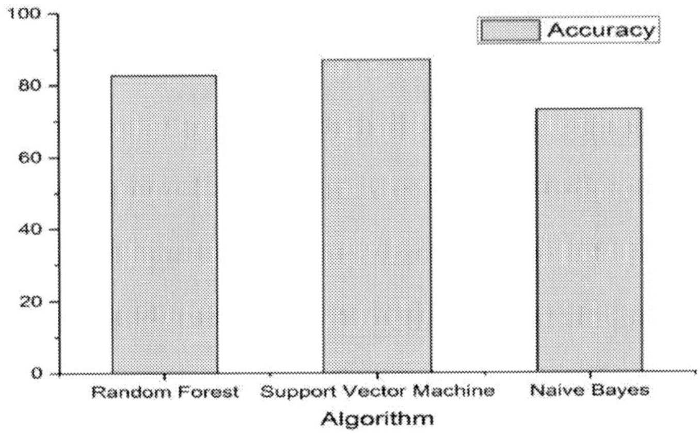

Figure 2. Comparison of classification algorithms in terms of accuracy.

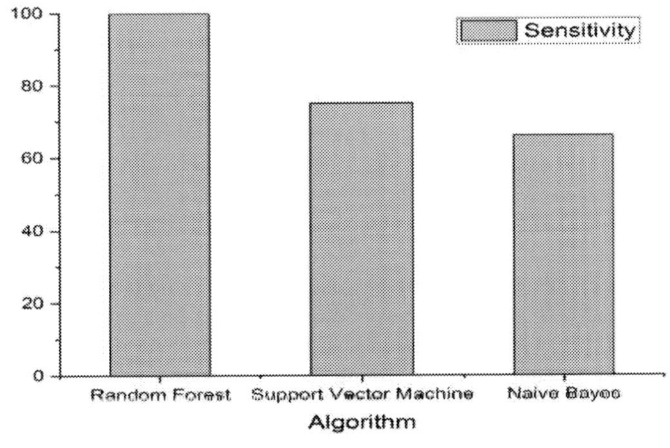

Figure 3. Comparison of classification algorithms in terms of sensitivity.

Figure 2 and figure 3 respectively show the accuracy and sensitivity of these three techniques discussed in this section.

Conclusion

Since the existence of COVID-19, the learning of the students has drastically changed. In this review article, we discussed this change in their learning through the dataset of the students. The dataset of the various students is collected and then pre-processed. The data is exhibited to the classification strategies by the various ML techniques like Support Vector Machine (SVM), Random Forest (RF) and Naive Bayes (NB). The analysis shows that accuracy is the best for support vector machine (SVM) method while sensitivity is best for random forest (RF) method.

References

[1] Anand, R., Sindhwani, N., & Saini, A. (2021). *Emerging Technologies for COVID-19.* Enabling Healthcare 4.0 for Pandemics: A Roadmap Using AI, Machine Learning, IoT and Cognitive Technologies, 163-188.

[2] Bao, W. (2020). COVID-19 and online teaching in higher education: A case study of Peking University. *Human Behavior and Emerging Technologies*, 2(2), 113-115.

[3] Zhang, W., Wang, Y., Yang, L., & Wang, C. (2020). Suspending classes without stopping learning: China's education emergency management policy in the COVID-19 outbreak. *Journal of risk and financial management*, 13(3), 55.

[4] Singh, H., Rehman, T. B., Gangadhar, C., Anand, R., Sindhwani, N., & Babu, M. (2021). Accuracy detection of coronary artery disease using machine learning algorithms. *Applied Nanoscience*, 1-7.

[5] Kirubasri, G., Sankar, S., Pandey, D., Pandey, B. K., Singh, H., & Anand, R. (2021, September). *A Recent Survey on 6G Vehicular Technology, Applications and Challenges.* In 2021 9th International Conference on Reliability, Infocom Technologies and Optimization (Trends and Future Directions) (ICRITO) (pp. 1-5). IEEE.

[6] Chatterjee, I., & Chakraborty, P. (2021). Use of information communication technology by medical educators amid covid-19 pandemic and beyond. *Journal of Educational Technology Systems*, 49(3), 310-324.

[7] Gupta, A., Anand, R., Pandey, D., Sindhwani, N., Wairya, S., Pandey, B. K., & Sharma, M. (2021). Prediction of Breast Cancer Using Extremely Randomized Clustering Forests (ERCF) Technique: Prediction of Breast Cancer. *International Journal of Distributed Systems and Technologies (IJDST)*, 12(4), 1-15.

[8] Bakshi, G., Shukla, R., Yadav, V., Dahiya, A., Anand, R., Sindhwani, N., & Singh, H. (2021). An Optimized Approach for Feature Extraction in Multi-Relational Statistical Learning. *Journal of Scientific and Industrial Research (JSIR),* 80(6), 537-542.

[9] Nash, C. (2020). Report on digital literacy in academic meetings during the 2020 COVID-19 lockdown. *Challenges*, 11(2), 20.

[10] Lassoued, Z., Alhendawi, M., & Bashitialshaaer, R. (2020). An exploratory study of the obstacles for achieving quality in distance learning during the COVID-19 pandemic. *Education Sciences*, 10(9), 232.

[11] Cao, W., Fang, Z., Hou, G., Han, M., Xu, X., Dong, J., & Zheng, J. (2020). The psychological impact of the COVID-19 epidemic on college students in China. *Psychiatry research*, 287, 112934.

[12] Jæger, M. M., & Blaabæk, E. H. (2020). Inequality in learning opportunities during COVID-19: Evidence from library takeout. *Research in Social Stratification and Mobility*, 68, 100524.

[13] Oliver, V. M., Hernández-Leo, D., & Rementeria, A. R. (2021, September). *Comparative analysis of interaction in online and in-person classes among undergraduate and Master students: Barriers and Facilitators*. In 2021 International Symposium on Computers in Education (SIIE) (pp. 1-6). IEEE.

[14] Jindal, M., Bajal, E., Singh, P., Diwakar, M., Arya, C., & Sharma, K. (2021, November). *Online education in COVID-19: Limitations and improvements*. In 2021 IEEE 8th Uttar Pradesh Section International Conference on Electrical, Electronics and Computer Engineering (UPCON) (pp. 1-5). IEEE.

[15] Li, Y., Song, Y., Matutu, S., & Li, D. (2021, October). *Research of an online-offline blending teaching model for the post COVID-19 era—Using a C programming language course as an example*. In 2021 IEEE Frontiers in Education Conference (FIE) (pp. 1-5). IEEE.

[16] Muñoz, G. M. H. (2020, December). *Collaborative Learning: From Face-to-face to the Virtual Classroom in the COVID-19 Era.* In 2020 X International Conference on Virtual Campus (JICV) (pp. 1-5). IEEE.

[17] Setiawan, A. W. (2021, April). *Utilization of Existing Resources to Support Active Distance Learning during COVID-19 Pandemic in Developing Country*. In 2021 IEEE Global Engineering Education Conference (EDUCON) (pp. 437-442). IEEE.

[18] Wattanakasiwich, P., Suree, N., Chamrat, S., Saengsuwan, W., Suttharangsee, W., Panrat, T., ... & Chantawannakul, P. (2021, October). *Investigating Challenges of Student Centered Learning in Thai Higher Education during the COVID-19 Pandemic*. In 2021 IEEE Frontiers in Education Conference (FIE) (pp. 1-7). IEEE.

[19] Tang, W., Cheng, Z., Lian, H., & Zhang, J. (2021, September). *Research on Online Teaching Satisfaction Based on Big Data and COVID-19.* In 2021 International Conference on Computer Network, Electronic and Automation (ICCNEA) (pp. 202-206). IEEE.

[20] Zhou, M., & Kalota, F. (2020, November). *Promoting Collaborative Learning through VR Technologies in the Era of COVID-19*. In 2020 Seventh International Conference on Information Technology Trends (ITT) (pp. 26-29). IEEE.

[21] Bouktif, S., & Manzoor, A. (2021, April). *Artificial Intelligence as a Gear to Preserve Effectiveness of Learning and Educational Systems in Pandemic Time*. In 2021 IEEE Global Engineering Education Conference (EDUCON) (pp. 1703-1711). IEEE.

[22] Hong, Y., Yang, J., Chen, Y., & Dong, H. (2021, January). *Research on the Development of Online Education in the Age of AI and 5G.* In 2021 2nd International Conference on Education, Knowledge and Information Management (ICEKIM) (pp. 233-237). IEEE.

[23] Kashif, M., Javed, M. K., & Pandey, D. (2020). A surge in cyber-crime during COVID-19. *Indonesian Journal of Social and Environmental Issues (IJSEI)*, 1(2), 48-52.

[24] Ayenew, B., & Pandey, D. (2020). Challenges and opportunities to tackle COVID-19 spread in Ethiopia. *Journal of PeerScientist*, 2(2), e1000014.

[25] Pandey, D., Pandey, B. K., & Wairya, S. (2021). Hybrid deep neural network with adaptive galactic swarm optimization for text extraction from scene images. *Soft Computing*, 25(2), 1563-1580.

[26] Khongsai, L., Anal, T. S. C., Rapheileng, A. S., Kh, T. S., Shah, M. K., & Pandey, D. (2021). Combating the spread of COVID-19 through community participation. *Global Social Welfare*, 8(2), 127-132.

[27] Pandey, D., Ogunmola, G. A., Enbeyle, W., Abdullahi, M., Pandey, B. K., & Pramanik, S. (2021). COVID-19: A framework for effective delivering of online classes during lockdown. *Human Arenas*, 1-15.

[28] Kassaw, C., & Pandey, D. (2021). The current mental health crisis of COVID-19 pandemic among communities living in Gedeo Zone Dilla, SNNP, Ethiopia, April 2020. *Journal of psychosocial rehabilitation and mental health*, 8(1), 5-9.

[29] Radwan, E., Radwan, A., Radwan, W., & Pandey, D. (2021). Perceived Stress among School Students in Distance Learning During the COVID-19 Pandemic in the Gaza Strip, Palestine. *Augmented Human Research*, 6(1), 1-13.

[30] Babu, S. Z. D., Pandey, D., & Sheik, I. (2020). Acts Of COVID19. *International Journal of Advanced Engineering*, 3, 2457-0397.

[31] Babu, S. Z. D., & Pandey, D. (2020). web-act of detecting covid 19. *Methodology*, 7(5).

[32] Nishitha, P., & Pandey, D. (2021). A Study on Student Perception Towards Online Education During COVID-19 Crisis. *Augmented Human Research*, 6(1), 1-7.

Chapter 13

COVID-19 Outbreak, Challenges and Possibilities for an Online System of Education

Iqbal Ahmad[1], Rani Gul[1,*] and Uzma Imtiaz[2]

[1]Department of Education, University of Malakand, Chakdara, Dir Lower, Khyber Pakhtunkhwa, Pakistan

[2]Department of English, FJWU, Rawalpindi

Abstract

Research has already underscored the importance of online teaching and learning and its many fold benefits for students. However, in the context of COVID-19 pandemic, many technological changes and challenges have occurred which have put educational institutions in a state of frenzy due to different issues related to online education and its effective implementation. This analytical review sheds lights on the technological changes and challenges to online education during covid-19 pandemic. The chapter further highlights how the new online education policy declarations by the governments in the new pandemic situation have triggered different reactions from student and teachers' communities and how the challenges could be overcome. It is becoming popular and challenging for educational institutions. Students and teachers have reported many problems and obstacles they faced during the during the COVID-19 pandemic, which need to be properly explored and addressed in order to successfully implement and ensure quality online education.

Keywords: online learning system, challenges and possibilities, covid-19 pandemic

---

* Corresponding Author's Email: dr.rani27@gmail.com.

In: An Interdisciplinary Approach in the Post-COVID-19 Pandemic Era
Editors: D. Pandey, B. Kumar Pandey, N. Sindhwani et al.
ISBN: 979-8-88697-052-4

1. Introduction

The effect and influence of information technology is recognized worldwide at all levels of education (Remuzzi & Remuzzi, 2020; Gul, R., Tahir., Ishfaq, U., Batool, T. 2021; Bukhari, S, K, S.; Said, Hamdan; Gul, R; Seraj, P, M, I. 2021). In the current developing scenario, the importance and contribution of information technology has increased tremendously due to the closures of educational institutions as a consequence of COVID-19 pandemic (Ahmad, I., Gul, R. 2021; Gul, R., Khilji, G. 2021). All this has raised challenges to students, parents, teachers and management alike. In the quarantined times, the role of information technology is helping people in solving the ongoing issues related to teaching and learning through innovative learning systems. It has provided enormous opportunities for teachers to use IT related resources in teaching and learning and evaluation upon completion of courses (Lippel, Sarma & Willcox, 2016),

There is an increased realization by all stakeholders like teacher, students and administrators for the maximum utilization of IT related sources across the globe due to the lockdown conditions. There is a general appreciation anacceptance of online learning as a suitable educational platform. The reason for the acceptability of IT related learning platform lies in the ease of learning, using learning kits, learning flexibility and supportive environment. However, despite all the benefits associated with the online learning, there are numerous challenges and limitations to online learning (Gul, R., Zakir, S., Ali, I., Karim, H., Hussain, R. 2021; Spina, Marrazzo, Migliari, Stucchi, Sforza & Fumagalli, 2020). These are to mention a few, social isolations, lack of face-to-face interactions of teachers and students, internet connection issues, learning accessibility issues and so on. In fact, online learning has not yet been accepted and adopted as a real learning platform as compared to traditional formal face to face learning. In the emerging pandemic situations, most of the educational institutions are attempting to make the use online learning easier for students and teachers to work out in the new normal time. For this purpose, efforts are being undertaken to explore different e-learning software for bringing optimal possibilities of ease and comfort to students and teachers.

As the pandemic is a recent development. Hence, teachers and students are learning to adapt to the new situation and new learning methodologies. Therefore, it has become even more important for all stakeholders to get to know the opinions of learners and educators to explore the level and satisfaction of learners' inclinations towards the novel method of teaching and learning. WHO (2020) reports that the outbreak of COVID-19 pandemic has changed the face of the world in all spheres of life, including education (Ahmad, I., Gul, R. & Zeb, M. 2022; Mukhtar, 2020). The pandemic was caused by a virus called corona SARS-COV-2 or simply corona. It originated from Wuhan, China and traveled fast to each nook and corner of the world, killing millions of people in the world (Ali, I., Gul, R., Khan, S. S., Karim, K. 2021; Schleicher, 2020). All educational institutions were closed down worldwide in the wake of the pandemic and online teaching and learning strategies were devised and implemented in the country's whole from pre-primary to higher education from April 2020 onward (Waris, Khan, Ali, Ali, & Baset, 2020; Adnan, M., & Anwar, K. (2020). Since then, the number of positive cases has increased to over 4 million, with over 7000 deaths. The government closed down the educational institutions again on 26 November due to the second wave of the pandemic, 2020 for the second phase (Shoaib & Abdullah, 2020). All higher education institutions, including colleges, were advised to follow standard operating procedures (SOPs) and social distancing to run the institutions with minimum essential staff and directed to switch on to online mode of teaching and learning. Since then, teachers and students have been found reporting different issues and problems related to the online mode of teaching and learning (Ayub, A., Gul, R., Ali, A., Rauf, B., M. 2021; Shahid & Zeb, 2020). In light of rising concern about the current COVID-19 pandemic, a growing number of universities across the world either postponed or canceled all campus events, such as workshops, conferences, sports (both intra and inter universities), and other activities. Universities moved rapidly to transition various courses and programs from face-to-face to online delivery mode (Gewin, 2020).

The review highlights the potential effects of the COVID-19 corona pandemic on the educational institutions, students and communities. To grapple with the challenges of the pandemic, many states of the world adopted plans and policies for educational institutions. As a result, travel to different countries was banned. Several health restrictions were imposed on educational institutions and protocols were devised to cope with the pandemic related issues and problems. People were advised to remain on self-isolation, quarantined in order to control the spread of the disease (WHO, 2020). Many states of the world closed down gyms, museums and film studios and other places of entertainment along with educational institutions. In the first reports, it was discovered that it only affected the elderly people. However, it also spreads to younger people later on and even children were found getting affected by the silent killer COVID-19 (Remuzzi A, 2020). Efforts to reduce the spread of the COVID-19 virus among the younger and adult populations have prompted the widespread closure of schools, colleges, universities, and other educational institutions in many countries. On March 25, 150, many countries of the world closed schools and educational institutions nationwide, affecting over 80% of the worlds. Many countries of the world used the local school protocols to close schools as a result of announcement related to the impact of COVID-19 (UNESCO, 2020). Much literature exists on the closures of educational centers and schools all around the world in order to control the further spread of the disease and recue the further chains of its transfer and transmissions (Kawano, Luca, Wheeler, 2009, 2010, 2018). This overview builds its arguments on four variables which are pandemic COVID-19, online education, technological changes and challenges and students' satisfaction.

The new virus appeared in December 2019 in the city of China named Wuhan (Cureus, 2020). The virus in a short period was spread to many places in the world, affecting people in large scale and causing deaths all around in no time. To decrease the degree of spread of the virus, many steps were taken by the governments worldwide (Xiang, 2020). Thus, in a short period, the whole of the world came in the grip of the virus and it became a world thread to human life (Spina, Marrazzo, 2020). The WHO announced the spread of the virus to a universal pandemic (WHO, 2020). In the March 2019, the virus spread to 177 countries in the world and, as a result, 732, 2435 people were affected due to the disease. Many regions in the world from Europe to Asia and Africa saw the outbreak of the problem

(Gul, R., Talat, M., Mumtaz, M., Shaheen, L. 2021; Khachfe, 2020). Due to this fast changing situation, the head of the WHO announced Europe as the epicenter of the virus (WHO, 2020). In China, some important steps were taken and, resultantly, the problem was reduced up to a great extent. However, in other parts of the world, the virus continued killing people on a large scale (Remuzzi, 2020)

The virus could not be completely controlled, and it attacked each time, with a new variant one after another. This resulted in the loss of many precious human lives and the whole geographic setting of the world underwent significant change. The outbreak forced people around the world to shift education from face-to-face mode to online mode (Gul, R., Ayub, A., Mazhar, S., Uddin, S., S., Khanum, M. 2021; Lippel et al., 2016). The online education was a new experience and practice for many teachers and students. Pitambar Paudel, (2020) stated that the COVID-19 changed the face of traditional classrooms and, as a result, education became more accessible as compared to the past. Now students could take classes anywhere from their homes. This experience provided a rich, flexible experience for students to learn. In this regard, the support of parents was highly important as they could facilitate the process as coach or guide and all this was not possible in the midst of their home related responsibilities. However, it was realized that online education presents a rich experience for learning along with different types of challenges (Bedford, Enria & Giesecke, 2020)..

The parental cooperation was a necessity in this regard, as even technology literate people face issues during implementing the process of online instruction. However, parents fully supported educational institutions with exceptions. The involvement of parents in the process of online education was a challenging task, as they were not familiar with the different processes involved in the system. The same for many teachers in many contexts due to less familiarity with the online teaching and the unavailability of required resources such as internet connectivity, internet signals and so on (Liu, Zhang & Chen, 2020). Truly speaking there were more issues for students especially from lowers grades. Many children could not cope with the online interaction and suffered to learn (UNESCO. 2020).

Literature shows that the familiarity of children with technical things starts as they grow old and go to senior classes. However, at the lower grades, they do not have experiences and technical knowhow of operating complicates systems like online education. Online education was a novel experience for them. They face problems for browsing through courses, connecting to online classes and so on. The process of online education was also challenging for the working parents. They could not manage the educational timetable of their children as it interfered with the work schedules. This added to their work related stress and anxiety in many cases (Kawano & Kakehashi, 2015). They tried their best to arrange the best type of learning environment for their children at homes, however, many parents could not do so. This led to lesser motivation by children towards online learning. Thus remaining at home without any proper learning environment became an lees interesting experience (Luca, Kerckhove, Coletti, Poletto, Bossuyt, Hens & Colizza, 2018).

2. Literature Review

The COVID-19 pandemic created a great disturbance in all spheres of life, including education at all levels, from primary to tertiary (Simamora 2020). The whole academic calendars in the world underwent changes as a result of the coronavirus. Majority of the schools, college and universities were closed down. Many people were self-quarantined (UNESCO, 2020). All the educational institutions were closed and faced challenging situations (Dhawan 2020). Especially, in higher education, students and lecturers could not meet in the classroom, and they reacted in different ways to the changing conditions. The educational institutions, in order to grapple with this situation, formulated alternative strategies such as accomplishing online learning. Digital learning helps in enhancing and improving the students' academic outcomes (Simamora 2020). Many educational institutions cancelled teaching and learning unless further notices However the new shifting from face to face to online learning created its own challenges and limitations, the main challenges for countries of the third world was the availability of digital infrastructure in remote and rural areas and lack of experience in teachers to use technology-based platforms for teaching. In due to course of time, technological and instructional challenges were highlighted as well as insights gained while teaching in the times of COVID-19 (Sunasee, 2020). This resulted in a widening of existing digital divide in the country (Hassan 2020). Some other issues included low internet speed, preventing picture, sound clarity which makes delivery of knowledge difficult for both the instructor and the student (Shukla 2020). The closure of educational institutes across the world imposed a bigger challenge for the education system. People from different reputed educational institutions developed a huge e-content and placed on the cloud for almost open access to the people across the globe. The world was literally taken under surprise by the COVID-19 and to ensure the health safety, almost all the institutions pertaining to teaching has been closed. People were compelled to adopt to online teaching-learning to compensate for the educational loss and if it continues, the online mode has to be a good alternative for the traditional teaching-learning processes (2020). These circumstances made the management of the educational institutions along with students, teachers and parents to realize the importance of combating the problem on an urgent basis (Dhawan, 2020). Most of the world communities were not ready for this situation of transmission of education from face to face to digital mode. It needed more preparations in terms of resources and facilities. Hence, both teachers and students faced enormous problems (Farooq, 2020). It was highly challenging for those educational institutions which did not have enough experience with practical digital education earlier (Lakbala, 2016). Thus, lack of proper arrangements led to issues in teaching and learning process for students, communities and educational institutions alike (Demuyakor, 2020). Despite the challenging situation, many practical steps were taken by the educational institutions to grapple with the situation for the time being to provide relief and help to students (Almazova, 2020)

3. COVID-19 and Online Teaching Learning

Online learning refers to offering an organized and well managed e-learning activities within a structured system like student enrolments, examinations, assignment, descriptions of courses, lesson planning, messaging, syllabus and developing course related materials. Digital instruction or online teaching and learning emerged as the sole choice for implementing online teaching and learning during the pandemic crisis. In the current situation, adoption of technology-based instruction is a not a new exercise for teachers and students. Research has explained that online instruction is an advantage for students (Ahmed, Hussain & Farid, 2018). Before the current pandemic situation and social distancing era, many educational institutions used a technology integrated approach for instruction and blended learning. However, since the closure of the educational institutions because of the pandemic crisis, the teaching and learning processes have been entirely performed through online system (Hebebci, Bertiz & Alan, 2020). In another study, Febriani et al. (2020) explained that online teaching and learning positively contributes towards students' development. For example, in an online mode of instruction, students get wider opportunities of interaction with new technologies and its application and people from different context. This helps enhance their social creative and interpersonal skills, which cannot be expected in a close lectured based classroom environment where they cannot work independently. Teachers can also find more space to engage students in various learning activities and also create and use innovative ways of instructions by using many new technology tools, social media and other learning applications for approaching students. This can also be applied as alternative media in the teaching and learning process when learning activities cannot be conducted through face-to-face interaction. Research has shown that in an online learning situation, students find better chances of academic and social growth as they have greater access to different alternatives sources of learning through websites worldwide (Dhawan, 2020).

The World Economic Forum (2020) has also underscored the importance of online instruction for students and teachers. Studies have further supported this claim that on an average basis, students retained 25-26% more materials while studying online as compared to mere 8-10% in a traditional classroom.

Research ash also proved that students learn faster in an online learning activity, because virtual learning required 40-60% lesser time for study than a traditional classroom (Adnan & Anwar, 2020). In an online learning situation, students are more independent as they are involved in a self-paced mode, they repeat and reread the material according to their own personal requirement, can skip over as well as accelerate through concepts they choose (Deming, Goldin, Katz & Yuchtman, 2015). Despite the numerous benefits associated with digital or online education, academic circles have already reported numerous obstacles facing students and teachers in an online teaching and learning situation (Assareh&Bidokht, 2011; Naveed et al., 2017). Scholars have referred to a number of factors to be taken into account for supporting the online learning activities for students and teachers such as economic self-sufficiency, accessibility to adequate infrastructure and facilities, availability of internet network, skill for implementing technological sources for teaching and learning (Almanthari, Maulina & Bruce, 2020). Researchers have pointed out some important determining factors which are essential for successfully achieving the goals of online instruction. These factors are home education sources, space availability, parental digital literacy, parental education and language of school instruction and parental school involvement (Sankar, Kalaichelvi, John, Menon, Elumalai, Alqahtani & Abumelha, 2020). Generally, it has been observed that in the pandemic related online teaching and learning situations, children coming low socio-economic backgrounds have remained disadvantaged due to non-availability of digital learning resources at homes and performed poorly. Research has showed that different demographic conditions, geographical and social background of students affected their learning performance during the pandemic situation. It was observed during the pandemic that students of lower grades with working parents having no guidance have received major setbacks. Such children of low economic background especially remain deprived of proper guidance and support from parents for online learning. When parents go to work no one is available to guide them through the online learning activity. Hence, students find it challenging to connect to internet and other digital learning facilities in rural areas or village (Hasani, Adnan, Sensuse & Suryono, 2020). Research has already indicated that most of the students who are enrolled in online courses are affected by different factors such as age and gender, earlier knowledge of computer literacy and different learning styles of teaching and learning. Studies have confirmed that many learners have access to internet facilities like computers and mobile phone. Theory of diffusion of innovations is the most relevant theory that is used for the study of technology adoptions in education. According to this theory, the perceptions of user of the technology immensely affect the effective adoption of technology and its relative advantages.

4. Technological Changes and Challenges

The COVID-19 pandemic has spread fast all over the world and does not seem to stop in the coming future. It has created different infection modes and so far 20 million infections have occurred and over 0.73 people have lost their lives. The epidemic has gravely affected all fields of life including education around the world. People are going into social isolations for avoiding further spread of the disease. This has formed new habits and activities through online modes (Wheeler, Erhart & Jehn, 2010). Being a necessary field, education has been greatly affected by the epidemic since its emergence. Students have faced hard times to go to schools and participate in collective activities. Thus, learning and other related activities are conducted online using blended learning approaches. This method has given educational institutions new venues of training and development by blending online and traditional methods of teaching and learning (Chahrour, Assi, Bejjani, Nasrallah, Salhab, Fares & Khachfe, 2020). The most favorable modes of education include online interactive activities, data storage and evaluations. However, online learning has its own challenges and limitations and to overcome this, teachers and students should have certain skills, experience and relevant knowledge to operate upon technology and pedagogical techniques for course completions. To make it more successful, the infrastructure of online learning must be synchronous, capable and secured for providing support to teachers and students' interactions, for storing data, and evaluating the effectiveness courses (Xiang, Yang, Li, Zhang, Zhang, Cheung & Ng, 2016).

The COVID-19 pandemic created many challenges for educational institutions. To cope with the situation and ensure maximum learning outcomes for students during the pandemic, different initiatives were started. Social distancing was found to be a useful method during the pandemic (Marabelli, Vaast & Li, 2021). The governments of many countries issued instructions to all educational institutions to transfer and shift all their academic activities to online teaching and learning modc immediately, where traditional mode of instruction prevailed. All face-to-face teachers and learning activities were closed down. Teachers were advised to carry instructions online and provide learning materials using digital tools. The newly implemented digital or online system was supposed to end in September, 2020 was further prolonged due to the spread of the pandemic and all semester and other examinations were cancelled. Numerous problems arose due to the new regulations of the government having internal and external repercussions for students and teachers (Al-Azzam, Elsalem & Gombedza, 2020).

The decisions of the government to further prolonging the implementation of the online teaching and learning activities over the current year further exacerbated the situation and the list of problems doubled for both teachers and students. Although the COVID-19 pandemic provided new opportunities for researchers and scholars all over the world to evaluate its various dimensions and its effects on educational institutions, hence, a significant body of knowledge is developing. However, in the Pakistani context, still more research is needed to study the problems and its long standing effects on education. Researchers have studied the educational, social and psychological implications of the pandemic crisis (Sintema, 2020a and Toquero, 2020) on education; Usak et al., 2020 on social and psychological). There has been some discussion on the educational implications of the crisis and the use of online instructional modalities and its implementation during the COVID-19 pandemic (e.g., See Sintema, 2020b on digital virtual class; Basilaia & Kvavadze, 2020 on online education; Naciri et al., 2020 on learning mobile; Mulenga & Marbán, 2020 about digital learning).

The COVID-19 forced all educational institutions to shift to an online mode of teaching and learning. E-learning is not new to learners, nor is distance learning. The closures of educational institutions have created several consequences for students, teachers and administrators. They were deprived of opportunities for growth and development. However, the online mode of teaching and leaning has resolved this problem more effectively by providing to different digital learning systems. Online learning may bring greater confidence, lead to reduction of stress and may enhance the concerns and empathies. In the new situation, students and teachers have faced many difficulties, such as accessibility to learning opportunities, material preparations for online learning and so on. Sometimes, teachers have found the online-learning platform more interactive for lesson sharing and using audio-visual aids and engaging in learning activities. Despite the numerous challenges involved in the teaching-learning process, research has already underscored the advantages of online learning, including convenience, learning flexibility, time flexibility in terms of places for learners and instructors. More preparation for learning among the handicapped students has been referred to as the most common outcomes of online teaching.

Studies have reported that the effectiveness of online learning is affected by different factors like brief interactions, discomforts in instruction without physical interaction, little skilled orientation. Other studies have identified some more limitations of e-learning, such as poor internet connectivity, shortage of online-learning infrastructure, lack of relevant training courses for students and teachers, and inadequate internet speed and signals.

Conclusion

The existing body of research clearly shows that teacher is a key factor in the successful implementation of the online learning in educational institutions. Hence, there is a dire need to enhance the capacities of teachers through intensive training and capacity building measures for online teaching readiness and enhancement of their essential technology related skills, knowledge and expertise that would enable them to play their roles effectively as a key partner in implementing the online teaching and learning and in the promotion of students' learning outcomes in all subjects. The analysis of literature on COVID-19 reveals that it is an ongoing, emerging situation. It has affected all departments of life, including education. Hence, the online system of instruction is the most suitable method in the present time to overcome the issues and challenges of teaching and learning. There is a need to mobilise resources and experience in carrying out online teaching and learning and especially in the underprivileged areas of the world where the availability of traditional courses or blended learning facilities is limited.

References

Adnan, M., and Anwar, K. 2020. Online Learning amid the COVID-19 Pandemic: Students' Perspectives. *Online Submission*, *2*(1), 45-51.

Ahmad, I., and Gul, R. 2021. Impact of Online Service-Learning on Civic and Social Justice Behavior of Undergraduate Laboratory-Based Graduates. *Human Arenas*. https://doi.org/10.1007/s42087-021-00244-9.

Ahmad, I., Gul, R., and Zeb, M. 2022. A Qualitative Inquiry of University Student's Experiences of Exam Stress and Its Effect on Their Academic Performance. *Hu Arenas*). https://doi.org/10.1007/s42087-022-00285-8.

Ahmed, M. U., Hussain, S., and Farid, S. 2018. Factors Influencing the Adoption of e-Learning 2018in an Open and Distance Learning Institution of Pakistan. *Electronic Journal of e-Learning*, *16*(2), 148-158.

Al-Azzam, N., Elsalem, L., and Gombedza, F. 2020. A cross-sectional study to determine factors affecting dental and medical students' preference for virtual learning during the COVID-19 outbreak. *Heliyon*, *6*(12), e05704.

Ali, I., Gul, R., Khan, S. S., and Karim, K. 2021. An Evaluative Study of English Contrastive Rhetoric in Pashtu Speaking Areas of Pakistan: A Case Study of District Swat. *LINGUISTICA ANTVERPIENSIA*, Volume 2021, Issue-1. pp. 2183 – 2203.

Almanthari, A., Maulina, S., and Bruce, S. 2020. Secondary school mathematics teachers' views on E-learning implementation barriers during the COVID-19 pandemic: the case of Indonesia. *Eurasia Journal of Mathematics, Science and Technology Education*, *16*(7), em1860.

Assareh, A., and Bidokht, M. H. 2011. Barriers to e-teaching and e-learning. *Procedia Computer Science, 3*, 791-795. https://doi.org/10.1016/j.procs.2010. 12.129.

Ayub, A., Gul, R., Ali, A., and Rauf, B. M. 2021. Cultural and Educational Stress: A Case Study Of Brahui Speaking ESL and EMI Periphery Students. *Asian EFL Journal.*, *28(2.3).* https://www.elejournals.com/asian-efl-monthly-editions/aej-monthly-edition- 2021/volume-28-issue-2-3-april-2021.

Basilaia, G., and Kvavadze, D. 2020. Transition to online education in schools during a SARS-CoV-2 coronavirus (COVID-19) pandemic in Georgia. *Pedagogical Research*, *5*(4).

Bedford, J., Enria, D., and Giesecke, J. 2020. COVID-19: towards controlling of a pandemic. *Lancet*. Accessed on Sep. 202010.1016/S0140-6736(20)30673-5 2020.

Bukhari, S. K. S., Said, Hamdan; Gul, R., and Seraj, P. M. I. 2021. Barriers to sustainability at Pakistan public universities and the way forward. *International Journal of Sustainability in Higher Education*, doi.org/10.1108/IJSHE-09-2020-0352.

Bukhari, S. K. U. S., Gul, R., Bashir, T., Zakir, S., and Javed, T. 2021. Exploring managerial skills of Pakistan Public Universities (PPUs)' middle managers for campus sustainability. *Journal of Sustainable Finance & Investment*, 1-19. doi: 10.1080/20430795.2021.1883985.

Chahrour, M., Assi, S., Bejjani, M., Nasrallah, A. A, Salhab, H., Fares, M. Y, Khachfe, H. H. 2020. A bibliometric analysis of COVID-19 research activity: a call for increased output. *Cureus*. 12:e7357. 10.7759/cureus.7357 2020.

Creswell, J. W. 2015. Revisiting mixed methods and advancing scientific practices. In *The Oxford handbook of multimethod and mixed methods research inquiry*.

Deming, D. J., Goldin, C., Katz, L. F., and Yuchtman, N. 2020. Can online learning bend the higher education cost curve?. *American Economic Review*, *105*(5), 496-501.2015

Dhawan, S.: Online learning: A panacea in the time of COVID-19 crisis. *Journal of Educational Technology Systems*, *49*(1), 5-22.

Febriani, S. R., Safutri, J. T., Yusnawati, Y., and Anasrudin, A. 2020. Development of Literacy in Islamic Education in the COVID-19 Era for Elementary School. *Khalifa: Journal of Islamic Education*, *4*(2), 79-96.

Gul, R., and Khilji, G. 2021. Exploring the need for a responsive school curriculum to cope with the COVID-19 pandemic in Pakistan. *Prospects*. **51**, pp. 503–522. https://doi.org/10.1007/s11125-020-09540-8.

Gul, R., Ayub, A., Mazhar, S., Uddin, S. S., and Khanum, M. 2021. Teachers' Perceptions on Students' Cultural and Linguistic Diversity and its Impact on their Approaches towards Culturally Teaching Practices. *TESOL International Journal*, 16 (3.2).

Gul, R., Tahir. Ishfaq, U., and Batool, T. 2021. Impact of Teachers Workload on their Time Management Skills at University Level. *Indian Journal of Economics and Business, 20(3).*

Gul, R., Talat, M., Mumtaz, M., and Shaheen, L. 2021. Does Intelligence Matters in Teaching? Exploring the Impact of Teachers Intelligence on Teaching Pedagogies of Secondary School Science Teachers. *Multicultural Education*, *7*(3). doi: 10.5281/zenodo.4647944.

Gul, R., Zakir, S., Ali, I., Karim, H., and Hussain, R. 2021. The Impact of Education on Business Opportunities for Women Entrepreneurs in Public & Private Television Advertisements in Pakistan. *Industrial Engineering & Management Systems*, 20 (2), pp.140-147 DOI: https://doi.org/10.7232/iems.2021.20.2.140.

Hasani, L. M., Adnan, H. R., Sensuse, D. I., and Suryono, R. R. 2020. Factors Affecting Student's Perceived Readiness on Abrupt Distance Learning Adoption: Indonesian Higher-Education Perspectives. In *2020 3rd International Conference on Computer and Informatics Engineering (IC2IE)* (pp. 286-292). IEEE.

Hebebci, M. T., Bertiz, Y., and Alan, S. 2020. Investigation of views of students and teachers on distance education practices during the Coronavirus (COVID-19) Pandemic. *International Journal of Technology in Education and Science (IJTES), 4*(4), 267-28.

Kawano, S., and Kakehashi, M. 2015. Substantial impact of school closure on the transmission dynamics during the pandemic flu H1N1-2009 in Oita, Japan. *PLoS One*, 10:e0144839. 10.1371/journal.pone.0144839.

Khachfe, H. H., Chahrour, M. A., Sammouri, J., Salhab, H. A., Eldeen Makki, B. Fares. 2020. An epidemiological study on COVID-19: a rapidly spreading disease. *Cureus*, 12:e7313. 10.7759/cureus.7313.

Lippel, P. H., Sarma, S. E., and Willcox, K. E. 2016. *Online education: a catalyst for higher education reforms report*, Massachusetts Institute of Technology, Cambridge, MA.

Liu, W., Zhang, Q., Chen, J. 2020. Detection of COVID-19 in children in early January 2020 in Wuhan, China. *N Engl J Med*. Accessed on Sep. 2020 to 10.1056/NEJMc2003717.

Luca, G. D., Kerckhove, K. V., Coletti, P., Poletto, C., Bossuyt, N., Hens, N., Colizza, V. The impact of regular school closure on seasonal influenza epidemics: a data-driven spatial transmission model for Belgium. *BMC Infect Dis*, 18:29. 10.1186/s12879-017-2934-3 2018.

Marabelli, M., Vaast, E., and Li, J. L. 2021. Preventing the digital scars of COVID-19. *European Journal of Information Systems*, 1-17.

Mukhtar, S. 2018. Preparedness and proactive infection control measures of Pakistan during COVID-19 pandemic outbreak. *Research in Social & Administrative Pharmacy*.

Mulenga, E. M., and Marbán, J. M. 2020. Is COVID-19 the gateway for digital learning in mathematics education?. *Contemporary Educational Technology*, *12* (2), ep269.

Naciri, A., Baba, M. A., Achbani, A., and Kharbach, A. 2020. Mobile learning in Higher education: Unavoidable alternative during COVID-19. *Aquademia*, *4* (1), ep20016.

Naveed, Q. N., Muhammed, A., Sanober, S., Qureshi, M. R. N., and Shah, A. 2017. Barriers Effecting Successful Implementation of E-Learning in Saudi Arabian Universities. *International Journal of Emerging Technologies in Learning*, *12*(6).

Park, J., and Park, M. 2016. Qualitative versus quantitative research methods: Discovery or justification? *Journal of Marketing Thought*, *3*(1), 1-8.

Remuzzi A, Remuzzi: COVID-19 and Italy: what next. *Lancet*, accessed on Oct. 1402020:10.1016/S0140-6736(20)30627-9 2020.

Sankar, J., Kalaichelvi, R., John, J., Menon, N., Elumalai, K., Alqahtani, M. and Abumelha, M. 2020. Factors Affecting the Quality of E-Learning During the COVID-19 Pandemic from the Perspective of Higher Education Students. *Journal of Information Technology Education: Research*, *19*(1), 731-753.

Schleicher, A. 2020. The impact of covid-19 on education insights from education at a glance. Retrieved from oecd. org website: https://www. oecd. org/education/the-impact-of-covid-19-on-education-insights-education-at-a-glance-2020. pdf.

Shahid, R., and Zeb, S. 2020. Second Wave of COVID-19 Pandemic: Its deleterious and mortal repercussion in Pakistan. *Journal of Rawalpindi Medical College*, *24*(4), 288-289.

Shoaib, M., and Abdullah, F. 2020. Risk Reduction of COVID-19 Pandemic in Pakistan. *Social work in public health*, *35*(7), 557-568.

Sintema, E. J. 2020. Effect of COVID-19 on the performance of grade 12 students:Implications for STEM education. *Eurasia Journal of Mathematics, Science and Technology Education*, *16*(7), em1851.

Spina, S., Marrazzo, F., Migliari, M., Stucchi, R., Sforza, A., and Fumagalli, R. 2020. The response of Milan's emergency medical system to the COVID-19 outbreak in Italy. *Lancet*. *395*, 49-50. 10.1016/s0140-6736(20)30493-1.

Toquero, C. M. 2020. Challenges and Opportunities for Higher Education Amid the COVID-19 Pandemic: The Philippine Context. *Pedagogical Research*, *5*(4).

UNESCO. COVID-19 Educational disruption and response. (2020). Accessed: October 25, 2020: https://en.unesco.org/themes/education-emergencies/coronavirus-school-closures. 2020.

Usak, M., Masalimova, A. R., Cherdymova, E. I., and Shaidullina, A. R. 2020. New playmaker in science education: COVID-19. *Journal of Baltic Science Education*, *19*(2), 180.

Waris, A., Khan, A. U., Ali, M., Ali, A., and Baset, A. 2020. COVID-19 outbreak: current scenario of Pakistan. *New Microbes and New Infections*, 100681.

Wheeler, C. C., Erhart, L. M., and Jehn, M. L. 2010. Effect of school closure on the incidence of influenza among school-age children in Arizona. *Public Health Rep*. *125*, 851-859. 10.1177/003335491012500612 2010.

WHO: Coronavirus disease (COVID-2019) situation reports. Accessed: October 10, 2020: https://www.who.int/docs/default-source/coronaviruse/situation-reports/20200314-sitrep-54- covid-19.pdf. 2020.

Xiang, Y. T., Yang, Y., Li, W., Zhang, L., Zhang, Q., Cheung, T. and Ng, C. H. 2016. Timely mental health care for the 2019 novel coronavirus outbreak is urgently needed. *Lancet Psychiatry*, *7*, 28-/S2215-0366(20)30046-8.

Chapter 14

COVID Pandemic: Redefining the Employment Ecosystem

Sanjive Saxena*
Jagan Institute of Management Studies, Rohini, New Delhi, India

Abstract

The CORONA pandemic jolted the world out of the blue. For what was once envisaged as a common virus, it quickly began to unleash its power and took the whole world in its grip. The business units too suffered from the ferocious impact of the CORONA pandemic. The government of the country implemented a complete lockdown forcing citizens to stay indoors and to adopt and embrace work from home culture. While some of the essential work force team members were allowed to move around freely others were not too lucky. Business units, sensing that CORONA is here to stay for a longer period than what was anticipated, they began to feel the impact of the lockdown as the work force was geographically placed and the issues pertaining to business sustainability and continuity began to surface. Further, the switch over from physical mode of operations to digital mode of operations has had its own limitations. For example, moving to digital means of carrying out business processes wherein the team member is stationed at home, the control exists gets transferred to from the project manager to *each* member of the project team. This makes the task of managing the team extremely difficult and getting the work done within the constraints of time and with no direct mechanism for controlling the team it generates different set of challenges *and which needs to be addressed* somehow or the other. Further, in remote operations certain other aspects need to be taken into consideration. For example, aspect needed to ensure that the home

* Corresponding Author's Email: sanjive.saxena@gmail.com.

In: An Interdisciplinary Approach in the Post-COVID-19 Pandemic Era
Editors: D. Pandey, B. Kumar Pandey, N. Sindhwani et al.
ISBN: 979-8-88697-052-4

workplace is provides unhindered work environment so that the concerned team member does productive work. Another crucial aspect which was relatively easy to manage, when physical mode of operations was in vogue, is the issue of decay of trust and adoption of unethical practices during the course of carrying out day to day business operations. In virtual world, as mentioned, the control exists with the person operating remotely, he can log off at will or can switch off the mike or video at his own leisure. In such scenario, the issue of trust and unethical practices need to be taken into consideration. This chapter addresses in detail supported by specific real life examples to address the employment eco-system prior to COVID pandemic and during the COVID pandemic. In particular the chapter takes into consideration the aspects such as work environment, human resource policies, organization culture and the like. This is due to the fact an eco-system is an amalgamation of several components each serving a specific role.

Keywords: COVID, employment, eco-system, organization, refurbish, remote, team workforce

1. Introduction

Business units exist to cater to the societal needs of their society [1], and this they achieve by creating economic value. However, the creation of economic value is attained through human beings by providing them with employment [2]. Thus, employment creation, generation, and directing it towards the goal of meeting the business objective of serving the needs of society requires a well-established employment eco-system [3].

Formulating and developing an employment eco-system requires a well-thought-out strategy catering to addressing the issues of sustainability, economic value creation, expansion of business operations, and the overall development of human work force resources to maximise their potential so that they are in a position to provide greater economic value to society [4]. All this translates to the fact that an employable eco-system exists in an environment which is conducive to addressing the needs of the workforce and the needs of the business environment while maintaining a balance between the demands of the workforce as well as the demands of market dynamics.

As markets evolve, meaning societal needs undergo changes, so too does the employment eco-system. Thus, the advent of the CORONA pandemic [5, 6, 7] forced business units to relook and redefine their employment eco-system.

This chapter is designed to address the employment eco-system during the COVID pandemic. However, before starting, it becomes imperative to compare the employment eco-system prior to the advent of COVID. For the sake of confidentiality, the name of the organisation is abbreviated as BU throughout the chapter.

2. Stakeholders of Employment Eco-System

In simple parlance, the term "stakeholders" means all those entities that directly or indirectly have a say in the functioning of the organisation or the activity that is undertaken to achieve a particular objective.

In the context of the employment eco-system, several stakeholders exist [8]. Some of the stakeholders are tangible, while some are intangible. Worth mentioning is the fact that context is in terms of the employment eco-system. Some of the stakeholders involved in the employment eco-system are:

i. *Organizational employees: These are the organization's direct stakeholders who are directly involved in the core operations of the business units [8].*For example, in the software development organization, the software developers who are responsible for developing the software are the business unit's direct or core stakeholders. On the other hand, the supporting employees are the admin people, the human resources personnel, and the like, who are the indirect stakeholders.

ii. *The clients: The clients are the core stakeholders for an organisation because they provide the necessary impetus for the business's continuity and growth [9].*

iii. *Workplace environment: One of the most important and critical factors in the employment eco-system.*The importance of this stakeholder arises from the fact that in the presence of an ambient work environment, the productivity of the work force and hence the productivity of the organisation is enhanced immensely [9]. On the other hand, an improper workplace environment ensures that productivity is low and hence the organisation is unable to utilise the full potential of the workforce team members. Worth mentioning is the fact that when the workplace environment is mentioned, it is referred to as the hygiene, the proper work space, including the components of adequate sunlight, lighting and other aspects which directly impact the functioning of an individual.

iv. *Organization Culture: This component of the employment eco-system is in charge of ensuring the organization's work environment [10].* Work atmosphere, in essence, refers to the process of developing a set of unwritten procedures and processes which are designed to provide an ambient atmosphere for the employees to work in a cordial manner. Organizational culture differs from workplace environment in that the latter is more tangible, such as an adequate work environment that is cleaned on a daily basis.On the other hand, the organisational culture refers to more of an intangible element attached to it. For example, employees are playing low volume music to relieve stress or one of the employees brings a box of sweets depending on the day he is comfortable.

v. *Organizational policies: Other important components of the employment eco-system are organisational policies [10].* These policies are responsible for ensuring the smooth running of the business operations in terms of employee benefits, in terms of travel and stay arrangements, in terms of sanctioning of leaves, in terms of approvals and the host of myriad operations which provide liveliness to the functioning of the organization.

vi. *Infrastructure: Another critical component of the employment eco-system.* In essence, this deals with the process of providing resources to the workforce team members so that they are able to carry out the core functions of the business units by using the infrastructural resources [10]. An example of infrastructural resources is the server room, wherein the servers are kept in an atmosphere of air conditioning and are accessible to one and all. Another example of infrastructure is the seating space and capacity wherein the project team members can sit comfortably.

Figure 1 below depicts the various stakeholders in an employment eco-system.

3. Pre-COVID Employment Eco-System

As mentioned in the introductory section, an employment eco-system is impacted by the environment in which it operates. Thus, prior to the advent of the CORONA pandemic, the employment eco-system was well structured and operated as a coherent unit amongst various interfaces that existed within the organisational units [12].

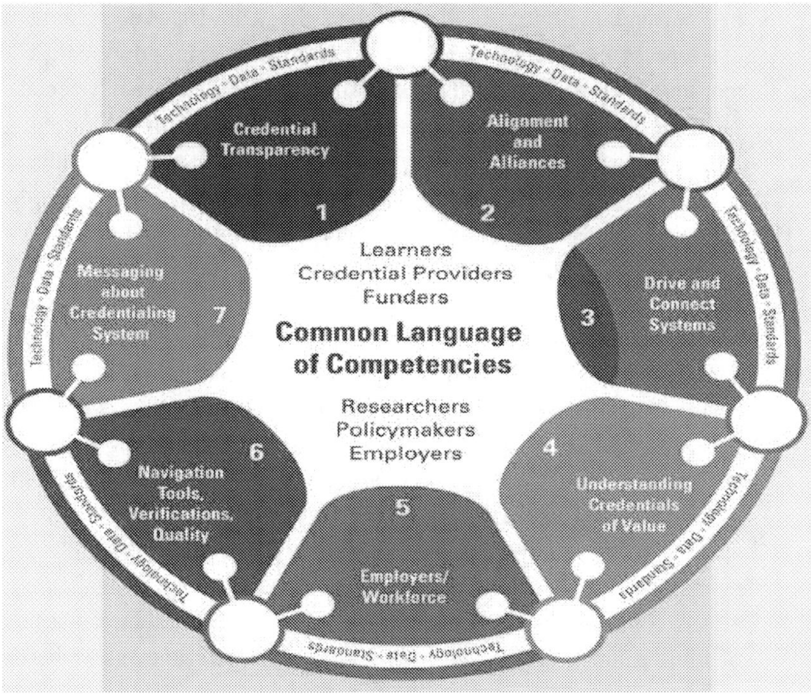

Figure 1. depiction of employee eco-system and its components.

The following points enumerate the highlights of the employment eco-system prior to the COVID pandemic.

1) The employment eco-system was centred on the basis of organisational design and the behaviour of the organisational units, which functioned in an integrated manner. Henceforth, this formed the crucial factor while identifying, constructing, and developing the employment eco-system [13]. This was achieved by means of a structured framework and process-oriented methodology in execution. Worth mentioning is the fact that the employment eco-system took into consideration various factors such as business model, revenue generation, geographical domain and the like. Let us take an example to illustrate with an example what is being discussed

in this context. An organization, BU, stationed in Mumbai, is into the development and execution of IT projects and IT products. The major business is generated from overseas operations, wherein BU has a significant physical presence. Depending on the business operations in a particular country, BU has opened physical offices depending on the type of operations it caters to. For example, BU only caters to marketing operations in the US, while in the UK it has opened a physical office which has departments such as Admin, Marketing, Human Resource, and Development, as the UK office needs several members of the workforce to carry out different IT projects for different clients. Thus, it is inferred that the employment eco-system is identified and developed on the basis of the business model, revenue generation, geographical domain, and other factors as applicable to the context in which BU is operating.

2) The employment eco-system functions on the basis of a well-established document system. The documented system covered various aspects such as organisation as well as department hierarchy and other aspects which covered the processes for dealing with conflicts and resolution and the means and mechanisms for addressing them in case of tailoring them for specific situations and needs. The relevant policies and documents were made available to all the concerned and affected workforce members either at the time of joining the organisation or in the case of specific events wherein they were necessarily needed to be referred to.
3) The employment eco-system was based on the concept of an effective and efficient system of management controls in place. This means that policies, processes, and procedures were drafted in such a manner which ensured that various organisational units operated effectively and efficiently by means of a measurement system which was modified on the basis of market dynamics [13], business requirements, and practical and feasible approaches. For example, in BU, in the case of leave approval for a worker, a well-defined process existed comprising of form filling and submitting to the appropriate authority, who in turn would forward the same to other authorities in question.
4) The employment eco-system policies and framework were based on the nature of the work on which the workforce members were required to work. In other words, the eco-system was designed and operated based on nature, the type, and other factors such as work

constraints, environmental challenges, and the like. For example, taking the case of BU, the members of marketing units, separate travel allowances along with the provision of waivers depending on the urgency and exigency of the work were in vogue. The same waiver was not applicable for workforce members from other units.

5) The employment eco-system was based on trust between the employee and the employer. In other words, there was a clear demarcation and conveyance of the specific tasks, duties, and other essentials that form a healthy work environment [14]. The basic premise is the fact that an employee functions best when both parties are true to their commitment.
6) The employment eco-system is responsible for ensuring a sanitary, safe, and clean working environment.This was one of the most important components for establishing a clean, hygienic, and safe work environment [15]. The basic premise of this concept is based on the belief that an employee or a workforce member is more productive in a clean, safe, and hygienic work environment than in the case of an unhygienic environment, which decreases productivity. Further, it is not only the work environment that forms the component of the eco-system, it is also the quantum of the work space allocated to the employees. In other words, space utilisation was also taken into consideration while developing the eco-system. For example, if an employee is seated in a cubicle, he or she is more productive than an employee who is seated in a hall wherein several members of the organisation are seated.
7) The employment eco-system operates on valuing human resources and looking after their basic necessities. This was another important concept which laid the foundation for the development, modification, and optimization of eco-system components. For example, BU valued human resources and had well-developed and formulated human resources policies in place. To take a specific example, if a workforce team member were to stay beyond office hours due to the exigency of the work, they would be compensated with a free meal. There was a policy in place that female work force team members were not required to stay past 8 p.m. and that BU would arrange for transportation to ensure they arrived home safely.
8) The employment eco-system was based on a well-structured, process-based system. In other words, while developing the eco-system, the focus was on designing the system for carrying out various day-to-

day operations by means of processes. This ensured that each of the workforce team members were aware of their specific roles, responsibilities, and authorities and the means of executing them so as to remain productive.

9) The employment eco-system was based on reward and motivation for the workforce team members. This was the most important aspect which provided thrust to the process of increasing the productivity of an employee. Human resource policies were identified and constructed to provide motivation and reward to the employee in the event of exceptional performance and superlative work. Separate policies and remuneration existed for rewarding teamwork and also individuals who had performed exemplary work.
10) The ecosystem was based on the career development and aspirations of individuals. The employment eco-system was based on the concept of career development and aspirations by providing them with training and an appropriate environment designed to ensure that they remain productive. Hence forth, regular training sessions, role plays, and other forms of events and activities are held to provide a conducive environment to enable them to break the monotony of the routine of working.
11) The induction process actually drives the employment eco-system. In other words, each of the functional heads were required to float their projected manpower requirements so that planning in terms of space capacity, in terms of finances, and in terms of job design was constructed so that operational functioning worked in a planned fashion.

From the above discussions, it is evident that the employment eco-system was well structured and thought of as a system comprising of several components, having clear, specific, and unambiguous roles and responsibilities for team members to carry out day-to-day operations *in a productive manner.*

Further, due to the physical proximity of the workforce (as they were stationed at the same physical office), demonstrable indicators were used to depict care, nurturing, and the demonstration of empathy for one another. Thus, the employment eco-system was centred on the fulcrum of trust, inclusiveness, and valuing each other so as to ensure that workforce team members remain productive as ever.

In order to illustrate the above points, let us take an example from the IT company. The name of the company is disguised as TBSoft, though the references mentioned are real, based on the industry experience of the author.

TBSoft is an IT company, based in Sector-62 Noida in the NCR region in India. The company is into development of Telecom software products. The clients of the TBSoft are government owned telecom service providers at domestic market whereas in the overseas operations the clients are the international telecom service providers. Since it has an international presence some of the workforce team members are stationed at international sites for different duration while some of the workforce team members are required to travel frequently for short duration.

The organizational structure of the TBSoft, has the CEO at its head, followed by Associate Vice President (AVP) and Chief Technological Officer (CTO). The CTO was responsible for Software and as such he was required to travel to various national and international sites. On the other hand the AVP was required to travel frequently to many countries for carrying out marketing related activities. The CTO was supported by several Project Managers who in turn was supported by software developers. In addition to this, there were various support staff members who main responsibility was to provide the ambient work environment for software developers. Table 1 depicts the various employment eco-system components which existed prior to CORONA pandemic

Table 1. Depiction of various employment eco-system components prior to COVID

Sr. No.	**Employment Eco-system Component**	**Practices followed for Implementation**
1	Human Resource Policies	Policies were designed to benefit and motivate the employee
		Policies were communicated to each employee for ensuring transparency
		Policies were designed to address the issues and grievances faced by the employees
		Policies existed which clearly defined the role, responsibility and authority of Human resource function

Table 1. (Continued)

Sr. No.	Employment Eco-system Component	Practices followed for Implementation
2	Organizational structure	Existence of well-structured framework for managing the business operations
		A clear hierarchy of the key stakeholders
		A clear and well developed inter-related processes for handing day to day operations
		Effective implementation of Organizational structure
3	Travel policies	A clear and specific policy for travel of employees
		Domestic travel policy components clearly conveyed to travellers
		International travellers policy different from domestic travel policy with clear instructions conveyed to travellers involving VISA, Forex and other documentation fee including stay arrangements to be borne by the company
		Separate travel components based on category and rank of the employee
4	Work environment	Clean and hygienic work environment
		Appropriate infrastructure so as to ensure that productivity work from employee is maintained
		Arrangement for travel by Cab and food in case of over stay by employee. Expenses to be borne by organization
		Cleanliness measures every week end such as pest control, sanitization etc.
5	Team composition	Balanced team composition comprising of experienced, semi experienced and new entrants
		Each team members role clearly specified to improve the productivity of the team
6	Infrastructure	Existence of network structure in terms of computers, server and necessary software for providing an ambient work environment
		Availability of sufficient quantity of lap top desktop and other tools needed for carrying out necessary work
		Security implementation measures

Thus, from the above Table 1 it is evident that the design of the employment eco-system is based on improving the work productivity of the organization.

4. COVID Pandemic Employment Eco-System

The advent of COVID completely shattered the existence of a stable, predictable, suitable, and well-accustomed employment eco-system [16, 17]. For this reason, the government imposed a complete lockdown across the country, thereby giving little time for planning and shifting of resources for continuity of business operations.

Initially, the employment eco-system tried to adjust to changed circumstances. However, as COVID is still raging across the world, the employment eco-system has been redefined for the sustainability of business operations and their continuity.

The following points enumerate the transformation of the employment eco-system triggered by the COVID pandemic.

1) Due to CORONA, the focal point of the employment eco-system was the adoption of digital technology for carrying out business operations [18, 19]. Hence, the imposition of lockdown and restrictions on movement, business continuity channels moved to digital media. Consequently, the employment eco-system was redefined to shift the focus from physical presence from office premises to virtual offices operating from home premises by using digital technology. On account of this, many of the employment-related policies were required to be redesigned and redrafted to accommodate this digital office in addition to the already existing physical office operations, as some of the essential workforce team members needed to physically come to the office.
2) Although digital technology adoption was forced on employees, it had far-reaching consequences for the employment eco-system.For example, some of the employees, due to location, faced connectivity issues which hampered their work, while some of the employees failed to work in a productive manner due to small children or otherwise. Thus, the employment ecosystem was faced with the dilemma of balancing the needs of the organisation as well as the needs of the employee. In other words, the employment eco-system was redrafted to provide an ambient work environment at a remote place of operations.
3) On the other hand, due to remote work culture, several issues began to resurface in the context of work culture and work ethics. For example, operating virtually provides the employee with the option

to disengage at any point of time or to adopt several forms of inappropriate and unethical practises and behaviour, such as muting the mouse or deliberately citing technical issues when important things are being discussed. In such a scenario, the employment eco-system was redefined to address these issues in the form of several practical-based policies, such as keeping the video on during the office hours, showing up in formal attire, and the like.

4) One of the major issues which drove the need for redesigning the employment eco-system included the aspects of FOMO (fear of missing out). Perhaps this was the greatest problem that managers struggled to meet. During the physical mode of operations, this FOMO was simply non-existent due to the fact that team meetings were conducted in conference rooms and each of the team members had the feeling of being there. Due to remote or virtual operations, team meetings are being held regularly, yet the feeling of being missed out is clearly resurfacing and the team meeting gets reduced to 1-1 meetings. In such a scenario, employment eco-system components are redesigned to address these issues.

5) Due to the work from home culture being in vogue, in team meetings with the client or otherwise, most of the employees felt ashamed to switch on the video camera due to the condition of the room or the space in which they were required to carry out the work assigned to them by their managers. This created a sense of inferiority or inferiority in dealing with the business prospect and created a negative image in the client. the employment eco-system needed redesigning and redrafting of the components of the employment eco-system. Consequently, strategies are drawn up to address these issues.

6) On the other hand, the workforce is the engine of the business, and it is this workforce that must be recruited on a regular basis.And the process of recruitment commences with the brief profile of the human being. Consequently, whatever was earlier known as resume or bio data in a physical format is now switched to video format wherein the prospective candidates are shortlisted on the basis of the video profile and evaluated for further processing. In other words, the contactless medium of interface has become the norm.

7) The employees working at home need to be kept motivated so as to overcome the humdrum work and monotonous work environment wherein the concept of 9 to 5 work timings simply does not exist. The workforce team members are overloaded with work, and official

meetings are usually conducted at odd hours. In other words, most organisations have started to address this issue of scoping the work timings of their employees.

8) The COVID pandemic wreaked havoc across the world in the form of many people getting impacted. However, organisations function on the basic premise of valuing and protecting their workforce employees from the aftermath of COVID impact. Hence, organisations have come up with the strategy of ensuring that their work force members are safe and protected while remaining productive at work.
9) As work from home became the norm, the trust factor was tested in the most stringent manner due to the fact that controlling the team members remained a challenge. Thus, some of the employees began to adopt malpractices, which affected the business operations.
10) As the work from home culture gained traction, significant changes are being observed in the workforce team members. These changes have necessitated the need for redefining the employment eco-system.

Table 2. Depiction of various employment eco-system components during COVID

Sr. No.	Employment Eco-system Component	Practices followed for Implementation
1	Human Resource Policies	Human resource policies are redefined to meet the exigencies posed by CORONA pandemic
		Refurbishing of home work environment for certain category of employees such as marketing operations. This includes redesigning of home work space so as to generate the impression of working from physical office premises
		Enhanced travel allowance and other benefits for certain category of work force team members such as news reporters visiting containment zones
		Allowances such as LTA are redesigned to provide the benefits to employees into some other form since travel is not the option in near future
		Redesigning of work related policies such as no calls from office after office hours

Table 2. (Continued)

Sr. No.	Employment Eco-system Component	Practices followed for Implementation
2	Organizational structure	Redesigning of organizational structure to ensure that it is able to adjust to hybrid mode of working
		Redesigning of organization structure with respect to operational processes such as removing the layered approach for approvals and the like
		Redesigning, renaming and restructuring certain designations so as to meet the business objectives without any compromise on the productivity of business operations
3	Work Environment	Provision for redesigning the home work environment by redrafting policies designed to benefit the employee working from home
		Provision for ensuring that employees take adequate precautions such as release from work for getting vaccinated
4	Team Composition	Provision to form the team based on redefined policy based on situational circumstances such as network connectivity, home environment such as disturbance arising due to small kids and the like
5	Employee Engagement	Separate policies formulated, redesigned and implemented to keep the employee engaged with the aim to prevent employee from adopting unethical practices such as switching off video or mike.
		Acquiring tools and software designed to determine whether the employee is actually working or has just logged in for the terminal
6	Organizational Culture	Redesigning policies to provide more thrust to the organizational working culture in terms of greater commitment to trust and adherence to ethical practices as the employee is working in virtual manner
		As the virtual means of work connectedness has become the norm, the organizational culture pertaining to meeting codes and etiquettes are conveyed to team members so as to ensure that professionalism is observed. For example, the team members forgets to switch off the mike after he has given his speech
7	Employee preference to work in hybrid mode	As COVID pandemic rages across the world and employee are somewhat stressed being confined at home, several organizations have redefined their strategy to provide hybrid mode of working. This ensures that employee stays in touch and does not fall prey to FOMO symptom

Table 2 above indicates the various employment eco-system components during COVID.

Conclusion

This chapter has covered the various aspects pertaining to the impact of COVID on the employment eco-system. An employment eco-system is critical to an organization's long-term viability and continuity of operations.However, certain moot questions need to be addressed, and answers must be obtained. Thus, the following questions must be answered. The questions are: what is an ecosystem? What are the constituents of an eco-system in the context of employment? What are the drivers of an eco-system? What means and mechanisms are needed to ensure that the eco-system is sustainable?

In seeking answers to these questions, the chapter tried to provide the solution in the context of employment. Further, it is seen from the above context and discussion that COVID has redefined the boundaries in terms of formulation of new components, formulation of new mechanisms for management and controlling of these new components, and how these components are integrated to work with one another.

We have discussed many aspects of the impact of COVID on employment. However, it is observed that there has been a complete redefinition of the meaning of the term "employment." What was earlier understood as a safe harbour for carrying out functional operations is redefined to function under more strict terms and conditions.

In particular, even what is being observed today as a part of employment strategy will give way to new business modalities. For example, leading business units have asked employees to refurnish their work space at home so as to reflect the impression that the employee is working from office premises. This indicates that work from home is here to stay and the leading business units are keen to ensure that employees work from home. Consequently, we may see two or more than two employment policies existing in the organization; one for those who are willing to work from home; one for those who are working from physical premises and the other for those who are working in hybrid mode.

Further, as a part of this employment strategy, some of the leading companies have already implemented this concept based on the distance they have from the office premises and consequently their pay package and their work terms and conditions have been redefined. Further, in addendum, some companies have redefined their employment contract to include a portion of the compensation package to be given to employee who are coming to office as they are risking their life so that some of the employee work from home and stay safe.

References

[1] Porter, M. E., & Kramer, M. R. (2011). *Creating shared value: Redefining capitalism and the role of the corporation in society.* Harvard Business Review, 89(1/2), 62-77.

[2] Lackéus, M. (2018). *"What is Value?"–A framework for analyzing and facilitating entrepreneurial value creation.* Uniped, 41(1), 10-28.

[3] Nicholas, D. B., Mitchell, W., Dudley, C., Clarke, M., & Zulla, R. (2018). An ecosystem approach to employment and autism spectrum disorder. *Journal of autism and developmental disorders*, 48(1), 264-275.

[4] Hilborn, Ray, (2007), *Defining success in fisheries and conflicts in objectives, Marine Policy*, Volume 31, Issue 2, Pages 153-158, ISSN 0308-597X, https://doi.org/10.1016/j.marpol.2006.05.014.

[5] Anand, R., Sindhwani, N., & Saini, A. (2021). Emerging Technologies for COVID-19. *Enabling Healthcare 4.0 for Pandemics: A Roadmap Using AI, Machine Learning, IoT and Cognitive Technologies*, 163-188.

[6] Meivel, S., Sindhwani, N., Anand, R., Pandey, D., Alnuaim, A. A., Altheneyan, A. S., ... & Lelisho, M. E. (2022). Mask Detection and Social Distance Identification Using Internet of Things and Faster R-CNN Algorithm. *Computational Intelligence and Neuroscience, 2022.*

[7] Khongsai, L., Anal, T. S., AS, R., Shah, M., & Pandey, D. (2021). Combating the spread of COVID-19 through community participation. *Global Social Welfare*, *8*(2), 127-132.

[8] Farooq, K., Yusliza, M. Y., Wahyuningtyas, R., Haque, A. U., Muhammad, Z., & Saputra, J. (2021). Exploring challenges and solutions in performing employee ecological behaviour for a sustainable workplace. *Sustainability*, *13*(17), 9665.

[9] Vázquez-Maguirre, M. (2018). Sustainable ecosystems through indigenous social enterprises. In *Entrepreneurial, Innovative and Sustainable Ecosystems* (pp. 173-189). Springer, Cham.

[10] Kassinis, G., & Vafeas, N. (2006). Stakeholder pressures and environmental performance. *Academy of Management Journal*, *49*(1), 145-159.

[11] Source: https://www.luminafoundation.org/campaign/learn-and-work-ecosystem/

[12] Kadam, S. A., & Pandey, D. (2020). Post COVID-19 Opportunities and Challenges for Micro, Small and Medium Enterprises (MSMEs) Development in India. *Asian Journal of Advances in Research*, 1-4.

[13] Babu, S. Z. D., Pandey, D., & Sheik, I. (2020). Acts of COVID19. *International Journal of Advanced Engineering*, *3*, 2457-0397.

[14] Juneja, S., Juneja, A., & Anand, R. (2019, April). Reliability modeling for embedded system environment compared to available software reliability growth models. In *2019 International Conference on Automation, Computational and Technology Management (ICACTM)* (pp. 379-382). IEEE.

[15] Singh, H., Rehman, T. B., Gangadhar, C., Anand, R., Sindhwani, N., & Babu, M. (2021). Accuracy detection of coronary artery disease using machine learning algorithms. *Applied Nanoscience*, 1-7.

[16] Pandey, D., Babu, S. Z. D., Noibi, T. O., Pandey, B. K., Dey, P. T. D., Kassaw, C., ... & Canete, J. J. O. (2020). A study on novel covid-19 epidemic threat for universal physical situation in 2020. *Asian Journal of Advances in Research*, 27-33.

[17] Jain, S., Sindhwani, N., Anand, R., & Kannan, R. (2022). COVID Detection Using Chest X-Ray and Transfer Learning. In *International Conference on Intelligent Systems Design and Applications* (pp. 933-943). Springer, Cham.

[18] Kohli, L., Saurabh, M., Bhatia, I., Sindhwani, N., & Vijh, M. (2021). Design and Development of Modular and Multifunctional UAV with Amphibious Landing, Processing and Surround Sense Module. *Unmanned Aerial Vehicles for Internet of Things (IoT) Concepts, Techniques, and Applications*, 207-230.

[19] Kirubasri, G., Sankar, S., Pandey, D., Pandey, B. K., Singh, H., & Anand, R. (2021, September). A Recent Survey on 6G Vehicular Technology, Applications and Challenges. In *2021 9th International Conference on Reliability, Infocom Technologies and Optimization (Trends and Future Directions)(ICRITO)* (pp. 1-5). IEEE.

Chapter 15

The Impact of Digital Transformation on Story Writing Skills in the COVID Era

Amalia Kusmalasari[1,*],
Widya Karmila Sari Achmad[2,†], Rohana[3,‡],
Binay Kumar Pandey[4,§], Digvijay Pandey[5,‖]
and Pankaj Dadheech[6,¶]

[1,2,3]Program Studi PGSD FIP Universitas Negeri Makassar, Indonsesia
[4]Department of Information Technology, College of Technology,
Govind Ballabh Pant University of Agriculture and Technology Pantnagar, India
[5]Department of Technical Education, IET, Dr. A.P.J. Abdul Kalam Technical University, Uttar Pradesh, India
[6]Department of Computer Science & Engineering (NBA Accredited),
Swami Keshvanand Institute of Technology, Management & Gramothan (SKIT),
Jaipur, Rajasthan, India

Abstract

This pre-experimental analysis sought to determine the impact of Web-centered teaching in Google Classroom upon the story-writing skills of 5th grade students at YPS Singkole Elementary School in East Luwu Township. One such study appears to be online learning with Google Classroom Focused, as well as 28 students in 5th grade at YPS Singkole

* Author's Email: amlisari97@gmail.com.
† Author's Email: wkarmila73@unm.ac.id.
‡ Author's Email: rohana@unm.ac.id.
§ Author's Email: binaydece@gmail.com.
‖ Corresponding Author's Email: digit11011989@gmail.com.
¶ Author's Email: pankaj@skit.ac.in.

In: An Interdisciplinary Approach in the Post-COVID-19 Pandemic Era
Editors: D. Pandey, B. Kumar Pandey, N. Sindhwani et al.
ISBN: 979-8-88697-052-4

Elementary School in East Luwu District. A data analysis of the data obtained by administering scriptwriting exams with a pre and posttest on the instructional concept of "My Living Place," sub theme 2: the variety of living organisms in my environment, was conducted. Descriptive analytics and inferential analysis are two data analysis tools. Depending on the outcomes of the inference statistical methods, the Wilcoxon test acquires a value of just 0.0001 or less than 0.05. It is reasonable to conclude that Web-based learning with Google Classroom Focus appears to have had a significant impact on the story writing abilities of the 5th grade classroom at YPS Singkole Elementary School in East Luwu District.

Keywords: learning, online, story, writing skills

1. Introduction

The development of science and technology is currently taking place so rapidly.

All the advances in science and technology seek to make it simpler for humans to accomplish tasks. The interests supported by adequate and quality human resources. The development of redundant advanced technology also influences the progress of education, despite the fact that this could help to improve human wealth efficiency. The advancements in interactive media and computer technology, as well as the use of the internet as a teaching tool, have also resulted in significant changes in existing instruction [1]. Digital learning in education helps to make it easier for educators and learners to cater to the process of learning, particularly for learners who struggle to comprehend the lecture in and out of the lecture hall. Learning outside of the classroom can be done independently by utilizing various media [2, 3].

In the early of 2020, there was an outbreak of the corona virus (COVID-19) across the country. More than 91% of the world's student population has been affected by school shutdown due to the COVID-19 Pandemic [4, 5, 6, 7, 8]. Learning at school was stopped and turned into learning at home. This also applies in Indonesia. Since March 2020 students are studying online from home. All academics, schools and teachers must think about student learning models at home so that learning material can be conveyed. The national education system can ensure that all students have equal access to quality education during this unprecedented crisis [9, 10, 11].

The Education ministry and Culture made the decision via Circular Number 4 of 2020, on Gaining Knowledge from Residence via Online Classes, that educators have been mandated to continue providing constructive learning opportunities for learners, without even being encumbered by requirements to accomplish all or most of the accomplishments of a graduate from college and graduate school coursework, but instead concentrating on skill sets teaching, such as on the Pandemic COVID-19. This offers a variety of work and learning activities from residence for classmates based on their interests but also circumstances, even while taking into account discrepancies in connectivity/learning amenities from residence and commenting on substantiation or goods of active learning from residence that really are quantifiable and beneficial for educators, without even the need to offer a quantifiable score/significance.

According to [12], in response to changes in the learning system, teachers must: (1) Implement the design and implementation of distance learning from face-to-face learning to distance learning in the midst of an unprecedented crisis, (2) be creative, innovative, adaptive, and flexible with good learning principles. (3) Build personal and professional strengths, acknowledge challenges and seek to overcome them.

SD YPS Singkole is a private elementary school under the Soroako Education Foundation. This school also implemented distance learning during the COVID-19 pandemic that hit Indonesia. This is because East Luwu Regency, especially Sorowako, is in the red zone of the COVID-19 pandemic. Based on the author's preliminary observations (online) in grade V SD YPS Singkole, the school has implemented online learning based on Google Classroom to all students. Google Classroom provides a productive teaching and learning activities by simplifying assignments and maintaining communication between teachers and students. The teacher creates a class by providing learning material, giving assignments, Student Worksheets (LKPD), Daily Test (UH), and Final Semester Assessment (PAS), and providing input to students.

The use of online learning based on Google Classroom will make it easier for students in distance learning because on the class assignment page students can share information about assignments, questions and learning materials. Students will easily view the task on the assignment page or in the class calendar and all class materials are automatically saved in the google drive folder. In addition, teachers can also send announcements, start live class discussions, quickly see students who have or have not completed assignments or answers.

In the Primary School curriculum placing Indonesian as the main subject. Learning Indonesian in elementary school includes four aspects, namely listening skills, speaking skills, reading skills and writing skills. Writing skills as one of the aspects of language skills at the final stage are mastered by students, because students can write well if a series of stages of aspects of language skills have been mastered by students. So it is expected that in the end students can meet the competency standards of language skills in the aspect of writing, namely writing effectively and efficiently of various types of essays in various contexts.

According to [13], writing is lowering or depicting letter symbols that describe a language that is understood by someone, so that other people can read the symbols of these letters. Writing effectively and efficiently requires a long series of processes. This process will be undertaken by students through the stages in language learning. One of the stages in advanced writing skills is writing stories. Writing stories has so far been considered a heavy burden for students, because it is considered difficult. Since infancy, children have begun to be good at telling stories or talking, such as talking about everyday events. This indicates that the child has the ability to express the contents of his thoughts through language orally.

Learning to write stories is included in the fifth grade elementary school curriculum. The purpose of learning to write stories in elementary school is so that students are able to write stories using the steps to write stories and pay attention to the correct use of spelling and punctuation. In accordance with the current situation and the government's appeal to conduct distance learning, the fifth grade teacher at SD YPS Singkole implements the learning process using Google Classroom.

Google Classroom application can be used to learn how to write a story. The teacher can provide material about the steps and how to write stories using suitable methods for distance learning. The teacher also provides LKPD with the guideline on how to make story outline of making a story outline according to the agreed theme, giving student learning outcomes tests in the form of writing stories from the developed story framework. In addition, the use of Google Classroom in story writing material as distance learning will still reach the cognitive, affective and psychomotor domains of students as well as start making it simpler for learners to study at home in accordance with learning goals. As a result, the accomplishment of the Google Classroom-based online learning platform is dependent on a multitude of factors, including schoolchildren, educators, educational materials, and information systems.

Such elements have been linked together in order to accomplish instructional outcomes.

Based on the description above, the researcher conducted a study entitled "The Effect of Google Classroom-Based Online Learning on Writing Stories for Class V Students at SD YPS Singkole, East Luwu Regency."

2. Online Learning

The use of technologies for teaching actions in Indonesian educational institutions has always been rising in response to the impact of online instructional strategies. The word "online" seems to be an abbreviation for "internet." As a result, digital training has become one of the e-learning methodologies and has been accomplished through a network connection. One such virtual education system was created by the Ministry of Education and Culture of the Republic of Indonesia through the Distance Learning Program (PJJ). PJJ seems to be a federally funded programmers that reaches out to schoolchildren on a national level (Ministry of Education and Culture of the Republic of Indonesia, 2014: 1

Digital learning occurs from anywhere, with schoolchildren kept separate from instructors, and it makes use of a variety of trying to learn references such as information, communication, and technology as well as other media as per law Number 20 of 2003, Article 1 Paragraph 15.

The understanding of online learning from various points of view of experts that has been summarized by [14] as follows:

a. A form of systematically organized independent learning, where counseling, presenters of learning materials, and supervisors and monitoring of student success are carried out by a group of teaching staff who have different responsibilities.
b. Learning is carried out remotely using device aid.
c. An instructional method that includes conversations among students learning as well as the use of articulation as an implied form of interaction between educators and students.
d. An education system that does not require a teacher at a certain place, but it is possible to have meetings between teachers and students at certain times.
e. A technique of communicating wisdom, abilities, and behaviors which is controlled through the usage of notions as well as runs

(division of labor), basic structures, and comprehensive usage media, particularly in instructional product lines,

f. An instructional approach wherein the instructional process is independent of the education process, requiring interaction among teaching students and staff to be enabled via printouts, digital media, as well as other media.
g. This is a type of learning that involves numerous aspects of gaining knowledge at numerous tiers of schooling that take place without straightforward and/or constant oversight of mentors to schoolchildren in the very same place, and which necessitates a procedure of preparation, organizing, and tracking from such an educational institution, and also supplying a procedure of mentorship and set of instructions both in the shape of immediate (real conversation) as well as simulated conversation).

Based on the understanding of online learning above, it can be concluded that distance learning is an effort to overcome educational problems with limitations between the teacher (tutor) and students to meet face to face by conducting learning that separates the teaching staff from students with the help of printed and electronic media. Such as e-mails, video conferences, soft files contain material that student can access without limitations on time and geographic location.

2.1. The Benefits and Drawbacks of Web-Based Learning

The advantages of using web based learning as per [15] are:

1. Is an effective, fast and credible communication medium to deliver online material from teachers?
2. Covers a wide area.
3. Contestants are provided a proper visual representation of their own discussion.
4. Smaller or larger classes: classes no longer require physical manifestation because all of this should be constructed in Web-based applications.
5. Every time, anyplace: world-wide in scope as well as available from every place. Through using asynchronous channels such as email or online conversation, to which students have access 24 hours per day,

e-learning eliminates the duration and location restrictions associated with conventional school features.

6. Learning appears to be a social and cultural process. Children can learn from each other by sharing details.
7. Improved student learning: through the Internet, organizations will be able to focus more on delivering education / training programs. Accommodate the entire learning process as well as transactions. The material can be designed in a multimedia and dynamic manner. People who participated in education could indeed communicate with different virtual libraries worldwide as well as utilize them even though survey resources to understand better educational materials. To widen respondents' understanding of their own educational materials, educators could indeed rapidly add citations in the form of reviews, industry developments, and technology and science forecasts using numerous perspectives.

While the disadvantages of online learning systems are as follows:

1. The learning and teaching methodology is more focused on instruction than learning.
2. Students who lack elevated amounts of educational encouragement are more prone to failure.
3. It is not that every location has a connection to the web (this is related to problems with the availability of electricity, telephones, or computers).

2.2. Google Classroom Based Online Learning

According to [16], online learning (or in Indonesian, namely the Google Classroom) seems to be a blended learning model devoted to each and every other context of teaching and learning, with the goal of finding alternatives to challenges in constructing, being able to share, as well as categorizing each assigned task without using manuscript.

According to [17], Herman in Hammi et al. (2017), Google Classroom appears to be a system that allows thc formation of online teaching. Furthermore, Google Classroom could be used to distribute coursework, complete assignments, and perhaps even grade completed coursework.

According to [18], Google-Classroom would be established to enable teacher-student interplay throughout virtual worlds. Even though educators don't really gather their coursework on articles, the category layout that uses Google Classroom would be genuinely environment-conscious.

This seems to be consistent only with the viewpoint of [19], who stated that classrooms inside this implementation have been created to aid educators in creating and collecting paper-free coursework, such as time-saving characteristics like the ability to easily create duplicates of Google documents for every student. To keep the place organized, the class could also construct drive directories for every task, so each student could. In addition to assignments, the teacher can also throw an idea in the form of contemporary ideas to be discussed in the Google Classroom class and if in real class learning (at school) there is a discussion of material that has not been completed, (via video) it can be completed and continued on the Google discussion forum Classroom.

2.3. The Advantages and Disadvantages of Google Classroom

The advantages and disadvantages of Google Classroom as stated by [20]:

1. Advantages
   a. The available communication features make it easy for users to discuss directly.
   b. Easy assignment system.
   c. Archive Course feature that make it easy for students to relearn material that has been missed.
   d. Ad-free.
2. Disadvantages
   a. It was found that students had difficulty using the feature at the beginning of the learning transition through goggle classroom.
   b. Learning becomes less efficient when students do not have data packages.
   c. Uneven signal coverage.

3. Story Writing Skills

3.1. The Essence of Writing Skills

Writing is a language skill to express thoughts and ideas to others. Writing is a communication activity in the form of delivering written messages to other parties. Writing activities involve the element of the writer as the delivery of messages, or the content of writing, written channels or media, and readers as message recipients, [21, 22] is expressing thought or feelings in form of writing (such as composing, writing letters, etc.) with writing.

As per [23] states that writing is essentially conveying ideas or ideas and messages using graphic symbols (writing). The ideas or messages conveyed depend on the development and level of knowledge of the author. Meanwhile, [13] states that writing is depicting or depicting graphic symbols depicting a language that a person can understand.

The ability to write must be continuously nurtured, because writing activities involve the recording of knowledge. As a language skill, writing is a complex activity because writers are required to be able to compile and organize the contents of their writing. Writing has many mental, intellectual, and social benefits for a person. According to [24], writing can increase intelligence, develop initiative power, foster courage, and stimulate the willingness and ability to gather information.

Based on the above opinion, it can be concluded that writing is a process of changing the form of thoughts (feelings) into symbols (writing) as a communication activity using written language as a tool or media.

3.2. Types of Writing Skills

Writing skills can be classified based on two different points of view. One such viewpoint would be an action or behavior in having to carry out writing abilities and the trying to write item's outcomes. As according [13], categorization of writing ability premised on the second view point outcomes in some kind of a split of writing ingredients or five classifications (2013), namely (1) narrative, (2) exposition, (3) description, (4) argumentation, and (5) persuasion.

1. *Narrative* (narrative). Storytelling seems to be a written work designed to narrate a sequence of events or lived emotions that emerge over time. The storytelling line of the article's aim is to tell the audience or viewer what the writer knows or has witnessed. The storyline emphasizes the flow of time and the presence of dispute [25].
2. *Exposition.* Normal description, also known as visibility, is really a type of article which attempts to describe, define, or analyze a topic in order to widen one's thoughts and insight. The author makes an attempt to analyze and specific the event or issue in order to introduce an interpretation of reality. It is critical to provide complete and accurate information when trying to write an exposition. Presentation is a form of text that is commonly used to communicate science-based explanations, such as documents, dissertations, term papers, or newspapers and publications. If you want to write about the rules for playing football, how planes work, how to make tempeh, for example, then the exposition type is very appropriate to use. Exposition tries to explain or explain. Suggested [26] that "An exposition writer would say, I will tell you all these incidents and events and explain them so that you can understand them." From the description above, it can be understood that in order to write an exposition, the writer must have adequate knowledge of the object to be worked on.
3. *Description.* An explanation seems to be a written or verbal summary of an item, location, ambience, or scenario. An explanation writer anticipates his viewer to be ready to "see" that which he seems to see, "listen" to that which he gets to hear, "experience" that which he feels, and reach the same "outcome" as him via his composing. This implies that characterization seems to be the consequence of analysis through five human senses and therefore is communicated in phrases [27].
4. *Argument.* Argumentation seems to be an a kind of written work which seeks to persuade or impact the audience to recognize the author's ideas. Assertions attempt to persuade an audience. This technique of persuading audiences can indeed be accomplished by displaying information, proof, or rationale outcomes [25].
5. *Persuasion.* Persuasion seems to be an article which includes a provocative or enticing elaboration that really can pique a readers' interest and cause them to think in and comply the author's inherent

bias or overt appeals. Persuasion, in other words, offers with issue of affecting others with dialect.

3.3. The Essence of Story Writing Skills

The story in prose is the result of the author's concoction based on his views, the judge's interpretation of events that have occurred, or events that only take place in fantasy [13]. It was further explained that in a story there are two important elements, namely (1) something is to be told and (2) there is a way of telling. The thing that is told is related to the theme, while the way of telling is related to the way the author arranges the story (plot), displays the character (characterization), displays the place (setting), places himself in relation to the story (how to tell the story). A story should have story elements, namely: (1) character/character, (2) characterization, (3) point of view, (4) setting, (5) structure,

Of the two classifications that have been mentioned, the type of essay or writing referred to in this study is the type of narrative (based on the first classification) or the type of story (based on the second classification). The elements of the story stated above are used as a reference in assessing the work/fictional story in this study.

3.4. Characteristics of Story Writing Skills

According to [28], the characteristics of writing stories are as follows:

1. In the form of stories about human events or experiences.
2. The events or events that are conveyed can be in the form of mere imagination or a combination of the two. Although imaginative, the author should not create a story at will. It is imaginary, but also logical.
3. The author is not concerned with the causal relationship of the problems raised. Narrative works are usually subjective in nature, both the content and the language used are strongly influenced by the author.
4. The emergence of conflict or the formation of a plot is often closely related to character, theme, and even setting. In narrative stories the

existence of conflicts, characterizations, and the course of the plot are very important.

5. It has aesthetics because the content and delivery are literary in the form of fiction.
6. Emphasizes the chronological atmosphere, with the intention of giving meaning to an event. So that readers can learn lessons from the story.

3.5. Elements of Story Writing Skills

The formation of a narrative essay or story has several elements, among others; plot, characterization, setting, point of view, and selection of event details. According to [29], the elements in writing stories are as follows:

i. Themes, a central idea that forms the basis of writing or works of fiction. In other words, the theme is the subject of discussion or the idea on which a story is based.
ii. Setting, setting is the environment in which events occur, both place and time. A place can be a space or things around it. While time is in the form of hours, days, months, years, seasons or periods.
iii. Character or characterization, is a clear depiction of a person who is shown in a story in the form of a protagonist or an antagonist.
iv. Plot, is the structure of a series of events in a story that is structured as functional interrelations that also mark the parts in the whole fiction. Flow governs the actions associated with other sections. With flow, events can be linked with other events in the unity of time.
v. A point of view, or what is called a point of view, discusses where a story is seen, whether as the first person who states in the story as me, or someone else who acts as a storyteller. The point of view states the function of how the author acts as a narrator in a narrative (narrator) acts and takes direct part (participant), or as an observer of the object of all actions in the narrative (observer). The author uses perspective as a way to convey the actors described in the story.
vi. Message, also known as mandate, of course every author has a message, either express or imply ideas or feelings that he wants to convey to the reader.

3.6. Steps for Story Writing Skills

According to [28], the steps in writing the story are as follows:

1. Determine the theme or message to be conveyed.
2. Set target readers.
3. Designing the main events that will be displayed in the flowchart.
4. Divide the main event into beginning-development (climax) and ending sections.
5. Breaking down the main events into details of the events that support the story.
6. Arrange characters/characterizations, settings/setting, and view-points.

3.7. Assessment of Story Writing Skills

To assess students' essay results, 3 methods or approaches are commonly used, such as (1) holistic assessment, (2) analytic assessment, and (3) priority aspect assessment [30]. Each method has advantages and disadvantages. Their selection and use is determined by the purpose of the assessment.

i. Holistic assessment or measurement is the assessment of an essay which is based on the appearance of the text or essay as a whole (holistically), not on aspects of the essay such as content, organization, mechanical aspects and so on. The belief that it is held is that the whole does not equal the parts, on the contrary the analytical judgment holds that the whole is the same as the sum of all the parts.
ii. Analytical assessment is carried out by scoring aspects of the composition of the essay. Each aspect of the essay has its own score with different weights. One display essay has many values or scores, a number of these values are then summed and the average value is calculated.
iii. The priority aspect assessment is similar to a holistic assessment, in that it assigns one score to each student's essay. However, this one value only describes the quality of certain aspects of the essay that are considered the most important. Aspects referred to, for example the

development of ideas, the order of information arrangement and so on [31].

4. Research Methods

This study uses an experimental method with a quantitative approach. In [32], there is defined a quantitative research as a systematic process of collecting and analyzing information (data) for various purposes.

This study used a Pre-Experiment design without a comparison group or a single group design in the presence of a pretest and posttest or so-called pretest and postest group design. The measurement is first carried out before the experiment (O1), then given treatment (X) for a certain period of time, then the measurement is carried out a second time (O2). This design can be described as follows:

O1 X O_2 [33]

O1 = pretest value (before being given treatment)

X = treatment, in this case the use of classroom-based online learning on story writing skills.

O2 = Posttest value (after being given treatment)

The number of population members in this study was 28 students taken from grade V SD YPS Singkole, East Luwu Regency. The sampling method used in this research is nonprobability sampling with the sampling technique used is saturated sampling. This is done because the population is relatively small, less than 30 people. The total sample size is 28 students.

The data collection techniques used in this study was observation, story writing skills tests, and documentation. The story writing skill test was carried out twice in the form of a pretest and posttest. The pretest was carried out in order to know the students' initial understanding of the teaching material for writing stories being taught and *posttest* implemented after students were the treatment. Documentation is used to collect data and support the completeness of data in research, so that the data obtained is accurate, such as data relating to the number of students, an overview of the learning process, and the conditions of the student's learning environment.

Learning is carried out for three meetings. The first meeting as a pretest. The second to third meeting as treatment (action). The third meeting served as a post test. The instrument used in this study was a writing skill test. The writing skill test serves to measure the level of individual writing skills as a result of learning.

Data analysis techniques in quantitative research are statistical analysis techniques, namely descriptive statistical analysis and inferential statistical analysis. Descriptive statistics are describing the data on the acquisition of students' story writing skills such as frequency, average value (mean), middle value (median), frequent values (mode), standard deviation (standard deviation), lowest data value (minimum), and the highest value of data (maximum). Inferential statistical analysis is intended to test the research hypothesis, before testing the hypothesis the prerequisite data test is carried out using the data normality test and hypothesis testing. Normality testing data in this study used the Shapiro-Wilk Normality Test to determine whether the data obtained were normally distributed. The data on the results of story writing skills are said to be normally distributed if the X and Y values have a significant value greater than 0.05. Then the hypothesis test is used to see the difference in the results of the questionnaire before and after online learning using google classroom is applied as well as the significant difference from the average story writing skills, then the data is analyzed using test Wilcoxon. In this study, the Wilcoxon test was used to test the two-sample comparative hypothesis to measure the effect of Google Classroom-based learning on students' story writing skills before and after being treated. The testing criteria, namelyH0 is rejected if the Asymp.Sig. > 0.05 and H1 is accepted if Asymp.Sig. <0.05.

5. Overview of Online Learning Based on Google Classroom Class V SD YPS Singkole Students

The online learning process based on Google Classroom for fifth grade students at SD YPS Singkole with the theme of my living place, the sub-theme of the uniqueness of my living place took place very effectively. Online learning based on Google Classroom has a positive effect on improving students' story writing skills. This is evidenced by the observations made during the online learning process that went well through Google Classroom. Based on the results of the observations made, there are still some things that have not been done on the teacher and student observation sheets with the percentage of implementation on the teacher's observation sheet of 95% which is in the very effective category. While the student observation sheet resulted up to 91.25%. This achievement has not been accomplished 100% due to various unfavorable situations and conditions. However, based on the results of these observations, it can be concluded that online learning based on Google Classroom is very effective.

6. Overview of the results of the Fifth Grade Student Story Writing Skills at SD YPS Singkole

Students' story writing skills with Google Classroom-based online learning in grade V SD YPS Singkole have increased after being given treatment. This is based on the results of the pre-test and post-test. The pre-test which was carried out on November 19, 2020, showed that the mean value was 80.73 and the standard deviation was 5.97. Based on the frequency table, it is known that there are no students who are in the low and very low category, 16 people are in the medium category, and 12 people are in the high category. Based on the results of the descriptive analysis, it was concluded that the pre-test of students' story writing skills was in the medium category.

The results of the final test (posttest) which was conducted on November 26, 2020 showed that the mean value was 86.73 and the standard deviation was 5.35. Based on the frequency table, it is known that there are no students who are in the low and very low category, 7 people are in the medium category, and 21 students are in the high category. Based on the results of the descriptive analysis, it can be concluded that the post test is in the high category.

Students' story writing skills through online learning based on Google Classroom have increased after being given treatment. This is based on before giving Google Classroom-based online learning treatment, students' story writing skills are in the medium category with an average of 80.73 until after giving Google Classroom- based online learning treatment they are in the high category with an average of 86.73.

7. Overview of the Influence of Google Classroom-Based Online Learning on the Story Writing Skills of Grade V Students of SD YPS Singkole

Based on hypothesis testing with inferential statistics, it shows that there is a significant influence on students' story writing skills after treatment in the form of online learning based on Google Classroom. Hypothesis testing using the Wilcoxon test with the help of the IBM Statistic Version 21 program obtained the Asymp Sig value. (2-tailed) from the Wilcoxon test results of 0,000 the value is smaller than 0.05, which means that H0 is rejected and H1 is accepted. Comparison of the mean pre-test and post-test, it is known that the average post-test score is greater than the pre-test score, namely 86.73> 80.73. It can be concluded that there is an effect of online Learning based on Google Classroom on the story writing skills of Grade V students at SD YPS Singkole.

Conclusion and Suggestions

Conclusion

Based on the results of the research conducted, there can be concluded several things related to the questions on the problem formulation, including:

1. The description of online learning based on Google Classroom in fifth grade students of SD YPS Singkole takes place effectively with the percentage of learning implementation based on teacher and student observation sheets who are in the very effective category.
2. The description of students 'story writing skills after online learning based on Google Classroom was higher than students' story writing skills before online learning based on Google Classroom was implemented.
3. There is an effect of online learning based on Google Classroom on students' story writing skills. This is because there is a significant difference between the pre-test and post-test scores of students' story writing skills after online learning based on Google Classroom is implemented. This is based on the results of hypothesis testing with

the Wilcoxon test, the significance value is 0.000, this value is less than 0.05.

Suggestions

Based on the research results, the researchers suggest the following:

1. For the school to strive to improve students' writing skills by facilitating factors that can improve writing skills such as the availability of reading materials, libraries, and supporting programs such as the literacy movement [34].
2. For students, to be more enthusiastic in every learning process and always train they to foster interest in writing even though learning takes place online.
3. For other researchers, it can be used as a reference in conducting similar research and can develop this research with different variables with a wider population.

Funding Statement

The author(s) received no financial support for the research, authorship, and/or publication of this article.

Availability of Data and Material

The data samples have been taken from classroom.

Contributions

All authors approve the final manuscript.

Ethics Declarations

Ethics Approval

Not Applicable (as the results of studies does not involve any human or animal).

Consent to Participate

Not Applicable (as the results of studies does not involve any human or animal).

Consent for Publication

Not Applicable (as the results of studies does not involve any human or animal).

Conflict of Interests/Competing Interests

No conflict of interest from all authors

References

[1] Syamsuddin, K., Rohana, T. K., Taufiqurrakhman, M., Pandey, D., & Enbeyle, W. Impact of COVID 19 on Education in Indonesia and Interdisciplinary Issues.

[2] Parthiban, K., Pandey, D., & Pandey, B. K. (2021). Impact of SARS-CoV-2 in online education, predicting and contrasting mental stress of young students: a machine learning approach. *Augmented Human Research,* 6(1), 1-7.

[3] Babu, S. Z. D., Pandey, D., & Sheik, I. (2020). Acts Of COVID19. *International Journal of Advanced Engineering,* 3, 2457-0397.

[4] Anand, R., Sindhwani, N., & Saini, A. (2021). Emerging Technologies for COVID-19. *Enabling Healthcare 4.0 for Pandemics: A Roadmap Using AI, Machine Learning, IoT and Cognitive Technologies,* 163-188.

[5] Jain, S., Sindhwani, N., Anand, R., & Kannan, R. (2022). COVID Detection Using Chest X-Ray and Transfer Learning. In *International Conference on Intelligent Systems Design and Applications* (pp. 933-943). Springer, Cham.

[6] Kamalraj, R., Neelakandan, S., Kumar, M. R., Rao, V. C. S., Anand, R., & Singh, H. (2021). Interpretable filter based convolutional neural network (IF-CNN) for glucose prediction and classification using PD-SS algorithm. *Measurement,* 183, 109804.

[7] Meivel, S., Sindhwani, N., Anand, R., Pandey, D., Alnuaim, A. A., Altheneyan, A. S., ... & Lelisho, M. E. (2022). Mask Detection and Social Distance Identification Using Internet of Things and Faster R-CNN Algorithm. *Computational Intelligence and Neuroscience,* 2022.

[8] Jain, S., Kumar, M., Sindhwani, N., & Singh, P. (2021, September). SARS-Cov-2 detection using Deep Learning Techniques on the basis of Clinical Reports. In *2021 9th International Conference on Reliability, Infocom Technologies and Optimization (Trends and Future Directions) (ICRITO) (pp. 1-5). IEEE.*

[9] Kohli, L., Saurabh, M., Bhatia, I., Sindhwani, N., & Vijh, M. (2021). Design and Development of Modular and Multifunctional UAV with Amphibious Landing, Processing and Surround Sense Module. *Unmanned Aerial Vehicles for Internet of Things (IoT): Concepts, Techniques, and Applications,* 207-230.

[10] Juneja, S., Juneja, A., & Anand, R. (2020). Healthcare 4.0-Digitizing Healthcare Using Big Data for Performance Improvisation. *Journal of Computational and Theoretical Nanoscience,* 17(9-10), 4408-4410.

[11] Pandey, D., Aswari, A., Taufiqurakman, M., Khalim, A., & Azahrah, F. F. (2021). System Of Education Changes Due to COVID-19 Pandemic. *Asian Journal of Advances in Research,* 10-15.

[12] Kadam, S. A., & Pandey, D. (2020). Post COVID-19 Opportunities and Challenges Formicro Small Medium Enterprises (MSMES) Development in India. *Asian Journal of Advances in Research,* 1-4.

[13] Nishitha, P., & Pandey, D. (2021). A Study on Student Perception Towards Online Education During COVID-19 Crisis. *Augmented Human Research,* 6(1), 1-7.

[14] Zargar, S. A., Islam, T., Rehman, I. U., & Pandey, D. (2021). Use of Cluster Analysis To Monitor Novel Corona Virus (COVID-19) Infections In India. *Asian Journal of Advances in Medical Science,* 1-7.

[15] Sharma, M., Sharma, B., Gupta, A. K., Khosla, D., Goyal, S., & Pandey, D. (2021). A Study and Novel AI/ML-Based Framework to Detect COVID-19 Virus Using Smartphone Embedded Sensors. In *Sustainability Measures for COVID-19 Pandemic* (pp. 59-74). Springer, Singapore.

[16] Lelisho, M. E., Pandey, D., Alemu, B. D., Pandey, B. K., & Tareke, S. A. (2022). The Negative Impact of Social Media during COVID-19 Pandemic. *Trends in Psychology,* 1-20.

[17] Pandey, D., Islam, T., & Malik, M. A. (2021). Novel coronavirus disease (SARS-COV-2): an overview. *Asian Journal of Advances in Medical Science,* 8-12.

[18] Pandey, D., Gul, R., Canete, J. J. O., Rocha, I. C. N., Gowwrii, G., Pandey, B. K., & Peter, S. N. (2021). Mental Stress in Online Learning during the Pandemic: An Assessment of Learners' perception. *Asian Journal of Advances in Research,* 37-49.

[19] Pandey, D., Pandey, B. K., Noibi, T. O., Babu, S., Patra, P. M., Kassaw, C., ... & Canete, J. J. O. (2020). COVID-19: Unlock 1.0 risk, test, transmission, incubation

and infectious periods and reproduction of novel COVID-19 pandemic. *Asian Journal of Advances in Medical Science,* 23-28.

[20] Pandey, D., Babu, S. Z. D., Noibi, T. O., Pandey, B. K., Dey, P. T. D., Kassaw, C., ... & Canete, J. J. O. (2020). A study on novel covid-19 epidemic threat for universal physical situation in 2020. *Asian Journal of Advances in Research,* 27-33.

[21] Mora Murillo, M. F., Mora Murillo, W. A., & Pandey, D. (2020). Efficient Prediction and Analysis of the possible evolution of the SARS-CoV-2 in COVID-19 applying the SEIR model in Santo Domingo de los Tsáchilas, Ecuador. *Enfoque UTE,* 11(4), 101-115.

[22] Pandey, D., Islam, T., Magray, J. A., Gulzar, A., & Zargar, S. A. (2021). Use of statistical analysis to monitor novel coronavirus-19 cases in Jammu and Kashmir, India. *European Journal of Biological Research,* 11(3), 274-282.

[23] Hastuti, Mulyati. 2002. Strategi & Metode dalam Model Pembelajaran. [Strategies & Methods in Learning Model] Jakarta: Referensi (GP Press Group).

[24] Pandey, D., Ogunmola, G. A., Enbeyle, W., Abdullahi, M., Pandey, B. K., & Pramanik, S. (2022). COVID-19: A framework for effective delivering of online classes during lockdown. *Human Arenas,* 5(2), 322-336.

[25] Radwan, E., Radwan, A., Radwan, W., & Pandey, D. (2021). Prevalence of depression, anxiety and stress during the COVID-19 pandemic: a cross-sectional study among Palestinian students (10–18 years). *BMC psychology,* 9(1), 1-12.

[26] Khongsai, L., Anal, T. S., AS, R., Shah, M., & Pandey, D. (2021). Combating the spread of COVID-19 through community participation. *Global Social Welfare,* 8(2), 127-132.

[27] Yadeta, A. E., & Pandey, D. (2020). Analysis of the global impact of the pandemic (COVID-19) on construction industry: possible scenarios. *Current Trends in Civil & Structural Engineering,* 6(4), 1-8.

[28] Warsita, Bambang. 2011. *Pendidikan Jarak Jauh, Perencanaan, Pengembangan, Implementasi, Dan Evaluasi Diklat. [Distance Education, Planning, Development, Implementation, and Evaluation of Education and Training].* Bandung: PT.Remaja Rosdakarya.

[29] Rather, R. A., Islam, T., Rehman, I. U., & Pandey, D. (2021). Development of vaccine against coronavirus disease 2019 (COVID-19) In India. *Asian Journal of Advances in Medical Science,* 13-21.

[30] Meslie, Y., Enbeyle, W., Pandey, B. K., Pramanik, S., Pandey, D., Dadeech, P., ... & Saini, A. (2021). Machine intelligence-based trend analysis of COVID-19 for total daily confirmed cases in Asia and Africa. In *Methodologies and Applications of Computational Statistics for Machine Intelligence* (pp. 164-185). IGI Global.

[31] Kassaw, C., & Pandey, D. (2022). COVID-19 pandemic related to anxiety disorder among communities using public transport at Addis Ababa, Ethiopia, March 2020: cross-sectional study design. *Human Arenas,* 5(2), 312-321.

[32] Radwan, E., Radwan, A., Radwan, W., & Pandey, D. (2021). Perceived stress among school students in distance learning during the COVID-19 pandemic in the Gaza Strip, Palestine. *Augmented Human Research,* 6(1), 1-13.

[33] Kashif, M., Javed, M. K., & Pandey, D. (2020). A surge in cyber-crime during COVID-19. *Indonesian Journal of Social and Environmental Issues (IJSEI),* 1(2), 48-52.

[34] Ayenew, B., Pandey, D., Yitayew, M., Etana, D., Binay Kumar, P., & Verma, N. (2020). Risk for surge maternal mortality and morbidity during the ongoing corona virus pandemic. *Med Life Clin,* 2(1), 1012.

Chapter 16

Unlock: Boon or Bane for Mental Health?

Mariyam Abbas[1,*], Sarah Javed[1,†], Asif Hasan[1,‡] and Binay Kumar Pandey[3,§]

[1]Department of Psychology, Aligarh Muslim University, Aligarh, Uttar Pradesh, India
[2]Department of Information Technology, College of Technology, Govind Ballabh Pant University of Agriculture and Technology, Pantnagar, India

Abstract

Unlocking after lockdown in the current COVID-19 pandemic has apparently resulted in a new normal way of living, but its impact on mental health is still unexplored. The present study tried to explore the mental health status of the general population in India during the current unlocking phase of the COVID-19 outbreak. Furthermore, it aimed at examining the mental health burden in different age groups and finally to find out the association between psychological distress and education loss, financial loss, exposure and acquaintance of COVID 19. The study was conducted from Unlock phase 3.0 to Unlock phase 8.0 on 200 participants. A Google form was created and circulated on different online and social media platforms. The measures used were: the Impact of Event Scale (IES), Depression, Anxiety, and Stress Scale (DASS-21). The data was analyzed with the help of SPSS v21. Results clearly show that the mental health burden of the majority of the population fell into the normal category and a few into the mild category. As far as stress, anxiety, depression, and IES in the unlock phase are concerned, they still exist but in lower figures as compared to lockdown phases. Without a

* Corresponding Author's Email: smakaficionado786@gmail.com.
† Corresponding Author's Email: sarahjaved2015@gmail.com.
‡ Corresponding Author's Email: hasan7686@gmail.com.
§ Corresponding Author's Email: binaydece@gmail.com.

In: An Interdisciplinary Approach in the Post-COVID-19 Pandemic Era
Editors: D. Pandey, B. Kumar Pandey, N. Sindhwani et al.
ISBN: 979-8-88697-052-4

doubt, the burden on mental health has been greatly reduced, but it still exists among the general population.

Keywords: mental health, lockdown, unlock

Introduction

Noval Coronavirus (COVID-19) originated in Wuhan, the capital of China (Hubei province) in December 2019 and started spreading rapidly worldwide. In response, non-pharmaceutical interventions were promptly implemented in low- and middle-income countries due to limited healthcare capacity (Walker, Whittaker, Watson, Bagel in, Winskill, Hamlet, & Ghani, 2020). The Indian Government enforced a 3 week nationwide lockdown as a non-pharmaceutical intervention on March 24, 2020; on April 14, 2020 there was an extension of lockdown till May 3, 2020 followed by subsequent 2-week extended periods (Pandey, 2020) with substantial relaxations, which gave a massive blow to all sectors due to the strict lockdown and quarantine protocols as a preventive and emergency scheme to protect vulnerable or at-risk people (Chen, Zhou, Dong, Q, Gong, Han & Zhang, 2020). The United Nations and the WHO called India's response to the pandemic "comprehensive and robust," while addressing the lockdown restrictions as "aggressive but vital" for containing the spread of COVID-19 and building necessary health-care infrastructure.

The government of India started the unlocking phases, which involved restricting the functioning of certain areas. Except for pharmacies, hospitals, banks, grocery stores, and other essential services, all services and shops were closed; commercial and private establishments were closed—only working from home was permitted; all education and training were suspended; research institutions were closed; all places of worship were closed; all non-essential private and public transportation was suspended; and all social, political, sports, entertainment, academic, cultural, and religious activities were prohibited (St After passing through various unlock phases, presently, India is in unlock phase 8.0, which started on January 1, 2021 and continued till January 31, 2021 with several changes in limitations imposed on various sectors. Still, Indians are not enjoying total freedom and abiding to the new normal, that is, wearing fashionable masks to slow aerosol transmission (Khan, & Parab, 2020) and following precautions in a trendy manner to move out to limited open spaces.

The whole situation, from lockdown till the current unlock phases, has drastically changed the lives of people living in India for more than a year now. Though India has managed to put a rein on the count of COVID-19 cases, mental health (an individual's state of psychological and emotional well-being) has been dramatically affected (Pandey, 2021). Studies concerning previous outbreaks such as Ebola, swine flu or MERS have revealed that such situations cause a deep and wide range of negative impacts on the mental health of all kinds of people. An abundance of literature has clearly indicated the impact of COVID 19 on behavioral (Liu, Luo, Li, Hong, Chen, & Xia, 2020), emotional (Lima, de Medeiros Carvalho, Lima, de Oliveira Nunes, Saraiva, de Souza, & Neto, 2020) and psychological (Serafini, Parmigiani, Amerio, Aguglia, Sher, & Amore, 2020; Shi, Lu, Que, Huang, Liu, Ran, & Shi, 2020; Wang, Pan, Wan, Tan, Xu, McIntyre, & Ho, 2020) aspects of people. Research has shown an increase in depression, anxiety, stress (Rehman, Shahnawaz, Khan, Kharshiing, Khursheed, Gupta, & Uniyal, 2020), PTSD, and suicides (Thakur & Jain, 2020), not only in those with physical ailments (Dubey, Biswas, Ghosh, Chatterjee, Dubey, & Lavie, 2020), but also in the general population.

Though the unlock phases have brought little relief to mental health, which had severe negative consequences during lockdown phases all across the globe and so in India due to this unusual situation of health emergency and the social restrictions taken to control the COVID-19 spread (Wang, Pan, Wan, Tan, Xu, Ho, & Ho, 2020; Xiang, Yang, Li, Zhang, Zhang, Cheung, & Ng, 2020). Most of the studies done in India during unlock phases were focused on how many undetected classes of infected individuals are out there (Paul, Chatterjee, & Bairagi, 2020); residential mobility (Saha & Chouhan, 2021); how many lives partial unlocks can save (Shuler, Koukouvitis & Suematsu, 2020); but a few tried reexamining the mental health of people during the unlock phase.

Due to the dearth of cross-sectional literature and studies on the impact of unlocking phases on the Indian population's mental health, this study aims to fill the gap to provide essential information regarding the section of the general population whose mental health has been impacted drastically owing to COVID-19, which could procure evolving interventions and techniques to tackle the issue substantially so that people's social functioning after emerging out of this pandemic remains intact and gets better. This study will contribute to developing a mentally sound society in a crisis and, globally, may help fight unprecedented events more positively.

Keeping this in view, the objective set for the study was to explore the mental health status of the general population in India during the current unlock phase of the COVID-19 outbreak. The second one was to examine the mental health burden in different age groups and finally to find out the association between psychological distress and education loss, financial loss, exposure and acquaintance of COVID-19 patients.

Method

The study was conducted from unlock phase 3.0 (1st August, 2020 to 31st August, 2020) to unlock phase 6.0 (1st November, 2020 to 31st November, 2020). A Google form was created and circulated on different online and social media platforms such as WhatsApp groups, email ids, Facebook and LinkedIn. Online written consent was taken from all the participants before they answered the questions. The data was analysed with the help of SPSS v21. Descriptive analysis, t-test, and ANOVA were carried out to make inferences.

Sample

A total of 200 participants completed the survey. Of the total participants, 123 were males, 73 were females, and four preferred not to disclose their sex. The mean age of the participants was found to be 34.94. From the sample, 17 were adolescents, 107 were young adults, 22 were middle-aged adults, and 54 were aged over 55. 38 participants were married, 160 participants were unmarried, and 2 participants were widowed. The sample included 164 people from cities, 33 from rural areas, and only three from semi-urban areas. The sample consisted of 96 students. 28 were academicians, 20 were businessmen, 11 were engineers, 11 were housewives, 7 were doctors, 3 were lawyers, while the remaining 24 were marked as others (e.g., not employed, retired, etc.).

Measures

Demographics

Sociodemographic details were taken using a semi-structured questionnaire which included gender, age, marital status, place of residence, and employment status.

Other related questions were also asked, such as: Have you been exposed to a COVID-19 patient? You following the prescribed precautions for COVID 19? Has corona resulted in financial loss? Has Corona resulted in educational loss?

Response Accuracy Scale (Monaghan et al. 2020)

The scale is a one-item measure used to assess the accuracy of the responses placed at the end of the questionnaire. Participants indicated how accurately they responded to the questionnaire on a five-point custom scale ranging from "didn't read the questions in the survey at all" to "read all questions in the survey". Only respondents who chose to "read most parts of the survey" or "read all questions in the survey" were included in the study, and their responses were retained.

Impact of Event Scale (IES) (Horowitz, Wilner, & Alvarez, 1979)

The IES is a 4-point Likert-type scale (0 = not at all, 1 = rarely, 3 = sometimes, 5 = often) that assesses subjective distress resulting from a traumatic life event through 15 items. It contains two subscales: intrusion (intrusive thoughts, nightmares, intrusive feelings and imagery, dissociative-like re-experiencing) and avoidance (numbing of responsiveness, avoidance of feelings, situations and ideas). Each item was scored at 0, 1, 3 or 5, with the higher scores reflecting more stressful impacts. The scores for the intrusive subscale range from 0 to 35, and are the sum of the scores for items 1, 4, 5, 6, 0, 11, and 14. The scores for the avoidance subscale range from 0 to 40, and are the sum of the scores for items 2, 3, 7, 8, 9, 12, and 15. The sum of the two subscales is the total stress score. It is suggested that the cut-off point is 26, above which a moderate or severe impact is indicated. IES can be interpreted according to the following dimensions: 0 – 8 (Subclinical range), 9 – 25 (Mild range), 26 – 43 (Moderate range), and 44 + (Severe range). Cronbach's alphas were found to bc 0.89 for the IES scale.

Depression, Anxiety, and Stress Scale (DASS-21) (Lovibond and Lovibond 1995)

Mental health status was assessed using the DASS-21. It is a 2nd edition and shorter version of DASS-42 developed by Loviid, S.H. & Lovibond, P.F. (1995). This scale is composed of three subscales: depression, anxiety, and stress. Each subscale is composed of seven items, and each response was rated from 0 to 3, where 0 indicates "Did not apply to me" and 3 indicates "Applied to me most of the time". The depression subscale was assessed on items 3, 5, 10, 13, 16, 17, and 21. The total score depression subscale score was subdivided into normal (0–9), mild (1012), moderate (1320), severe (21–27), and extremely severe depression (28–42). Anxiety subscales were assessed on items 2, 4, 7, 9, 15, 19, and 20. The total score of the anxiety subscale was subdivided into normal (0–6), mild (7–9), moderate (10–14), severe (15–19), and extremely severe anxiety (20–42). The stress subscale is constructed by items 1, 6, 8, 11, 12, 14, and 18. The total score of the stress subscale was subdivided into normal (010), mild (11–18), moderate (19–26), severe (27–34), and extremely severe stress (35–42). Cronbach's alphas were found to be 0.88 for the stress subscale, 0.89 for anxiety, and 0.85 for depression subscales.

Data Screening

No missing values were found in the data. However, responses from eight people were excluded from the analysis as they did not choose option 4: "Read most parts of the survey" or option 5: "Read all questions in the survey carefully." of the response accuracy scale as prescribed by the authors. Thus, the final data comprised of 200 individuals (123 males and 73 females, with 4 preferred not to say). For analysis (age groups), all the respondents were taken into consideration.

Results

The average score of the participants on the revised impact of event scale (IES-R) questionnaire was 17.64 9.53. More than half of the participants (57%) had normal scores on the IES-R, but 39% had scores in the mild range. On the DASS, 63% had normal scores on the stress subscale, 59% on the anxiety subscale, and 54% on the depression subscale. Mild symptoms of stress were experienced by 39%, which is similar to the 35% who experienced mild symptoms of anxiety and the 34% who experienced mild symptoms of depression (Table 1).

Table 1. The mental health burden of the Coronavirus pandemic

	IES-R	**Stress**	**Anxiety**	**Depression**
Score, Mean± SD	17.64±9.53	9.30± 4.28	5.98±2.78	9.00± 4.74
Category, N (%)				
Normal	115 (57)	126 (63)	117 (59)	107 (54)
Mild	78 (39)	70 (35)	71 (35)	69 (34)
Moderate	05 (3)	03 (2)	11 (5)	18 (9)
Severe	2 (1)	01 (1)	01 (1)	6 (3)

Table 2. The mental health burden of the coronavirus pandemic among various age groups

Variable	Group	N	Mean	SD	Level	F	Sig
IES	Adolescents	17	16.41	10.6	Normal	3.04	.30
	Young Adults	107	19.44	8.2	Normal		
	Middle aged adults	22	14.27	9.6	Normal		
	Older aged adults	54	15.81	10.8	Normal		
Depression	Adolescents	17	8.70	2.68	Normal	18.38	<.001
	Young Adults	107	8.15	2.80	Normal		
	Middle aged adults	22	7.31	3.01	Normal		
	Older aged adults	54	12.55	5.69	Severe		
Anxiety	Adolescents	17	4.88	2.05	Normal	14.26	<.001
	Young Adults	107	7.00	2.60	Mild		
	Middle aged adults	22	5.81	3.40	Normal		
	Older aged adults	54	4.35	2.12	Normal		
Stress	Adolescents	17	7.17	3.69	Normal	2.43	.06
	Young Adults	107	9.40	4.64	Normal		
	Middle aged adults	22	10.50	6.16	Mild		
	Older aged adults	54	8.16	4.35	Normal		

Table 2 shows group differences in IES, stress, anxiety, and depression. On IES, the mean values were found to be 16.41, 19.44, 14.27, and 15.81 for adolescents, young adults, middle-aged adults, and older aged adults, respectively. Their respective standard deviations were found to be 10.6, 8.2, 9.6, and 10.8. No significant difference was found between the variables and all the groups fell into the normal category. In depression, the mean values were found to be 8.70, 8.15, 7.31, and 12.55 for adolescents, young adults, middle-aged adults, and older aged adults, respectively. Their respective standard deviations were found to be 2.68, 2.80, 3.01, and 5.69. A significant difference was observed (F = 18.38, p 0.01) among different age groups in depression. Adolescents, young adults, and middle-aged adults reported normal levels of depression, whereas older people reported severe levels of depression. On anxiety, the mean values for adolescents, young adults, middle-aged adults, and older-aged adults were found to be 4.88, 5.81, 4.35, and 3.40, with 2.05, 2.60, 3.40, and 2.12 standard deviations, respectively. A significant group difference was found in the anxiety scores (F = 14.26, p 0.01) among different age groups. Adolescents, middle-aged adults, and the elderly reported a normal level of anxiety, whereas young adults reported a mild level of anxiety. As far as the mean values on stress are concerned, adolescents, young adults, middle-aged adults, and older-aged adults reported 7.17, 9.40, 10.50, and 8.16, with 3.69, 4.64, 6.16, and 4.35 standard deviations, respectively. A non-significant difference in stress levels was also observed (F = 5.88, p.06). It was evident that the middle-aged population reported a mild level of stress, while adolescents, young adults, and the elderly population reported a normal level of stress.

Table 3. Educational and Financial loss faced, Experience of COVID 19 by family members, Contact history, precautionary measures used by the population during the epidemic

Variable	**Yes N (%)**	**No N (%)**	**Maybe N (%)**
Has Corona Resulted in Educational Loss	92 (46)	90 (45)	18 (9)
Has corona resulted in Financial Loss	118 (59)	65 (32)	17 (9)
Do you know someone among your friend or relative being a COVID 19 patient	110 (55)	86 (43)	4 (2)
Have you been exposed to COVID 19 Patient	32 (16)	97 (48)	71 (36)
Are you following prescribed precautions for COVID 19	173 (86)	15 (8)	12 (6)

The responses obtained to the 5 questions were in Yes, No, and Maybe. 46% of people reported that Corona has resulted in educational loss. More than 50% of the population said that Corona had resulted in financial loss. There were 55% of respondents who reported that their family or friends had been patients of COVID 19. The majority of the population (48%) were not exposed to COVID-19 patients. 86% of the population were following the prescribed precautions for COVID 19.

Table 4. Associations between Educational and Financial loss, contact history, precautionary measures, and the psychological impact during the pandemic

Variable	Yes	No	Maybe	F	Sig
Has Corona Resulted in Educational Loss					
IES	17.90±9.69	17.72±9.45	15.88±9.42	.340	.712
Depression	9.67±4.46	9.03±4.26	8.72±3.47	.684	.506
Anxiety	5.86±2.89	6.16±2.83	5.61±1.94	.429	.652
Stress	11.38±4.53	6.37±3.51	9.94±4.50	34.34	.000
Has corona resulted in Financial Loss					
IES	18.21±9.48	16.35±9.47	18.58±10.10	.887	.413
Depression	9.99±4.20	8.58±4.43	7.23±3.25	4.56	.012
Anxiety	6.16±2.78	5.70±2.88	5.76±2.46	.607	.546
Stress	9.82±4.57	7.70±4.74	8.23±4.93	4.56	.011
Do you know someone among your friend or relative being a COVID 19 patient					
IES	23.6±5.99	10.08±7.64	15.5±10.34	94.98	.000
Depression	9.19±4.08	9.43±4.63	9.50±2.08	.079	.924
Anxiety	6.40±2.74	5.45±2.78	5.50±2.88	2.95	.055
Stress	9.17±4.92	8.98±4.52	4.50±1.73	1.89	.153
Have you been exposed to COVID 19 Patient					
IES	18.62±9.98	14.24±9.32	21.83±7.79	15.03	.000
Depression	8.62V3.28	8.90±4.29	10.14±4.59	2.19	.114
Anxiety	5.93±2.86	5.85±3.01	6.16±2.44	.261	.770
Stress	9.31±5.47	8.52±4.13	9.50±5.14	.961	.384
Are you following prescribed precautions for COVID 19					
IES	17.76±9.51	14.33±9.26	19.91±9.95	1.26	.285
Depression	9.42±4.41	8.40±4.01	8.58±2.39	.572	.565
Anxiety	5.92±2.79	6.46±2.94	6.16±2.69	.287	.751
Stress	9.15±4.84	8.53±4.43	7.41±3.23	.828	.439

The associations between educational and financial loss, contact history, precautionary measures, and the psychological impact during the epidemic were assessed using one-way ANOVA. Participants who agreed with the statement that CORONA has resulted in education loss faced more stress (M = 11.38, SD = 4.53), followed by those who were in a dilemma (M = 9.9, SD = 4.50) and those who didn't agree (M = 6.27, SD = 3.51) with this statement. The difference was statistically significant at.01 (F = 43.34, p.001). Furthermore, participants who reported that CORONA caused financial loss had higher levels of depression (M = 9.99, SD = 4.20) and stress (M = 9.82, SD = 4.57), with significant differences (F = 4.56, p =.012). Participants who knew a COVID 19 patient among their friends or relatives had a greater impact on the event (M = 23.6, SD = 5.99) than those who did not have such relatives, and the difference was significant (F = 94.98, p =.001). Their anxiety level was also reported to be higher (M = 6.40, SD = 2.74) with a significant difference ($F = 2.95$, $p=.055$). Participants who had been exposed to COVID-19 patients had a significant difference in the impact of events ($F = 15.03$, $p =.001$), with those in dilemma having higher mean values ($M = 21.83$, $SD = 7.79$), followed by those who had been exposed ($M = 18.62$, $SD = 9.98$) and those who had not been exposed ($M = 14.24$, $SD = 9.32$). No significant association between participants' practising or not practising precautionary measures and mental health was seen among the participants.

Discussion

The main aim of the present research was to explore the mental health of the Indian population during the unlock phase of COVID-19. The data was collected from different age groups, including adolescents, young adults, middle-aged, and elderly adults for this study. The results clearly show that the mental health burden of the majority of the population was falling into the normal category and others into the mild category, and only a handful of people were having moderate and severe levels of mental health burden (Table1). This can be credited to the current phase of Unlock. We all know that India now has long come out of the lockdown restrictions imposed by the COVID-19 pandemic by introducing unlocking. The unlock phase has come along with new normal concepts that include social distancing, individual hygiene, mandatory mask wearing, isolation of infected or carriers, confinement at home for education or occupational activities, etc. The new normal concept limits the restrictions imposed during lockdown, which has ultimately resulted in a reduction of the mental burden. Another possible explanation for obtaining a normal range of mental burden in our study can be attributed to the passing of so many months of pandemic, which has made the general population-be it a child or an elderly person, or a normal person-accustomed to this life in a pandemic. And it's possible that a small percentage of people in the mild category exist because they're still battling the pandemic with uncertainty, using old methods, and having limited functioning according to unlocking norms.

As far as age-wise differences in stress, anxiety, depression and IES in the unlock phase is concerned, it still exists but in lower figures as compared to lockdown phases (Roy, Singh, Mishra, Chinnadurai, Mitra, & Bakshi, 2020; Xiong, Lipsitz, Nasri, Lui, Gill, Phan, Chen-Li, Iacobucci, Ho, Majeed, & McIntyre, 2020; Kene, 2020). When we specifically looked at different age categories in Table 2, results indicated that there was no significant impact of events or pandemics (IES scale score) and stress among all age group members now in unlock. However, in comparison to other age groups, middle-aged people showed a mild level of stress, which is significant because they have been overburdened by a lot of responsibility at home, i.e., trying to strike a balance to meet the needs of family members and job requirements—resulting in marital discord, family tension, or even a non-progressive approach to life—thus acting as a source of stress among the middle-aged.

However, a significant difference was found in the elderly for depression, indicating severe levels, where other age groups had normal levels of depression. This is majorly due to the fact that elderly people are considered the most vulnerable. They have been made trapped at home. Their way of utilising leisure time and other activities has been shackled considering the current scenario, which instead of the unlock norms hasn't reduced their depression level. Complementing studies like Swati Sharma (2020) did sentiment analysis during the unlock period and found people have high negative emotions. Even Mrugesh et. al., 2020 exclaimed through his work-the psychological uncertainty of when the pandemic end.

Additionally, there was a significant difference in anxiety among different age groups where young adults showed a mild level of stress. This is due to the fact that no matter unlock has brought relief in various aspects, education, finances, career, and most importantly time management, have always been a source of anxiety for young adults. Such negative emotions tend to intensify with the restrictions usually taken by the authorities to reduce the virus spread, such as closure of schools and businesses (Van Bortel, Basnayake, Wurie, Jambai, Koroma, Muana & Nellums, 2016) and this pandemic. It seems unpromising to college and university-going students of what the future beholds for them. The sudden shift to online learning systems and too much engagement on screen has brought so many new factors to adopt in order to function smoothly, which is causing worry among young adults (Al Najjar, Attar, Farahat, & Al Thaqafi, 2016).

Also, the findings of the current study shed light on whether mental health was associated with educational and financial loss during the unlock phase of COVID-19. Results revealed those who agreed with educational and financial loss that stress and depression were pretty evident. Our sample consisted of the majority of college and school-going students (N = 96). This pandemic was a major threat to them. Many unanswered questions regarding their future, career, job, and academics were continuously popping up in their minds and the repercussions of that were increasing their stress level. The current pandemic places demands on students and teachers alike to adapt to online teaching, albeit with limited resources, thus resulting in above-normal stress levels. Furthermore, because the majority of people in our sample came from urban areas (N = 169), where businessmen (N = 20), engineers (N = 11), lawyers (N = 7) and people in other non-essential services (N = 24) faced monetary loss due to lay-offs, job or business closures, pay deductions, and unlock tried to prevent the financial crises that the country was likely to face, academicians (N = 28) and doctors (N 7).

Furthermore, the findings reveal people who were exposed to COVID-19 patients or who knew someone among their friends or relatives was a COVID-19 patient regarded COVID-19 as a traumatic life event. These people scored highly on the IES scale. No doubt, COVID-19 has left an everlasting mark on the lives of many people. Many of them have lost their parents or siblings, and many watched them suffering alone in isolation. For those who couldn't do anything except pray, it might have had a huge mental impact. As losing someone from this new virus is dramatic for people, it is not easy to cope with the fact that flu-like symptoms are taking away or causing suffering in people's lives. No doubt our findings are in line with previous studies where it has been reported that the effects of COVID have an impact on the mental health of the general population. Studies concerning previous outbreaks such as Ebola, swine flu or MERS have revealed that such situations cause a deep and wide range of negative impacts on the mental health of all kinds of people. (Liu, Luo, Li, Li, Hong, Chen, & Xia, 2020; Lima, Carvalho, Lima, Nunes, Saraiva, de Souza, & Neto, 2020; Serafini, Parmigiani, Amerio, Aguglia, Sher, & Amore, 2020).

Conclusion

The present study is a frontrunner in exploring levels of anxiety, stress, and depression as well as posttraumatic stress symptoms in the Indian population during the unlocking phase. Our aim to assess mental health burden in the unlock phase has been assessed from a holistic lens and concludes that, no doubt, the burden has been reduced to a great extent, but still there is the existence of this burden among the general population. This reduction is to be attributed to the fact that people are accustomed to this change as more than a year has passed by. But still, the population is experiencing new, unpredictable, and rapidly evolving situations where they have to stay mostly confined at home; family dynamics have remarkably changed; and there has been a downfall in leisure activities and social life. The work situation has also changed markedly; many people have temporarily or permanently lost their jobs; some are working from home, sometimes with insufficient preparation for doing so; and those essential sectors are experiencing heavy workloads, increased levels of stress, and greater chances of exposure to the virus. The situation is still heavy instead of trying best by the government and citizens to curtail its impact as fear and concerns about the future continue to affect mental health.

In fact, in Wuhan, residents compared this health crisis with "the end of the world" (Lima, de Medeiros Carvalho, Lima, de Oliveira Nunes, Saraiva, de Souza & Neto 2020). Thus, the mental health burden exists among the general population during unlocking phases as well, in the form of depression among the elderly, as stress among the middle-aged, as anxiety among young adults, and posttraumatic symptoms among those who have been exposed to or know someone with corona.

Implication

To the best of our knowledge, no study has been conducted till date which has explored the mental health of the general population during the unlock phase of COVID 19. Hence, the findings of the study can be regarded as pioneering. The findings will assist education policymakers, mental health professionals, and other interested parties in developing better frameworks and interventions for unlock or lockdown phases, as unlock has been shown to be more successful in reducing mental burden than lockdown phases. Also, the findings may prepare the citizens to come out mentally sound once the situation is over and to utilise their potential to the maximum for themselves and the betterment of society.

Limitations

Although the research has made significant contributions and can be used by mental health providers to tackle the adverse psychological effects of COVID-19 during the unlock phase, It has some limitations. Firstly, the size of the samples in some groups was quite small. Secondly, the researchers utilised online Google forms for data collection that hindered the participation of a larger section of the population, such as those who do not have internet access, especially the underprivileged. Last but not the least, it was a cross-sectional study. Longitudinal study results would have given a clearer picture.

References

Al Najjar, N. S., Attar, L. M., Farahat, F. M., & Al Thaqafi, A. (2016). Psychobehavioural responses to the 2014 Middle East respiratory syndrome-novel corona virus [MERS CoV] among adults in two shopping malls in Jeddah, western Saudi Arabia. *EMHJ-Eastern Mediterranean Health Journal*, *22*(11), 817-823.

Chen, N., Zhou, M., Dong, X., Qu, J., Gong, F., Han, Y., ... & Zhang, L. (2020). Epidemiological and clinical characteristics of 99 cases of 2019 novel coronavirus pneumonia in Wuhan, China: a descriptive study. *The lancet*, *395*(10223), 507-513.

Dubey, S., Biswas, P., Ghosh, R., Chatterjee, S., Dubey, M. J., Chatterjee, S., ... & Lavie, C. J. (2020). Psychosocial impact of COVID-19. *Diabetes & Metabolic Syndrome: Clinical Research & Reviews*.

Hao, F., Tan, W., Jiang, L., Zhang, L., Zhao, X., Zou, Y., ... & Tran, B. (2020). Do psychiatric patients experience more psychiatric symptoms during COVID-19 pandemic and lockdown? A case-control study with service and research implications for immunopsychiatry. *Brain, behavior, and immunity.*

Kene P. (2020). Mental health implications of the COVID-19 pandemic in India. *Psychological trauma: theory, research, practice and policy, 12*(6), 585–587. https://doi.org/10.1037/tra0000750

Khan, M. M., & Parab, S. R. (2020). Simple Face Shield for Public as a Crucial Factor to Slow Aerosol Transmission During Unlock Phase of COVID Pandemic. *Indian Journal of Otolaryngology and Head & Neck Surgery*, 1-2.

Lima, C. K. T., de Medeiros Carvalho, P. M., Lima, I. D. A. S., de Oliveira Nunes, J. V. A., Saraiva, J. S., de Souza, R. I., ... & Neto, M. L. R. (2020). The emotional impact of Coronavirus 2019-nCoV (new Coronavirus disease). *Psychiatry research, 112915*

Liu, X., Luo, W. T., Li, Y., Li, C. N., Hong, Z. S., Chen, H. L., ... & Xia, J. Y. (2020). Psychological status and behavior changes of the public during the COVID-19 epidemic in China. *Infectious Diseases of Poverty, 9*(1), 1-11.

Lovibond, S. H., & Lovibond, P. F. (1995). *Manual for the Depression Anxiety Stress Scales* (2nd ed.). Sydney: DASS Publications.

Ma, Y. F., Li, W., Deng, H. B., Wang, L., Wang, Y., Wang, P. H., ... & Yang, Y. (2020). Prevalence of depression and its association with quality of life in clinically stable patients with COVID-19. *Journal of affective disorders, 275*, 145-148

Mandal, M., & Mandal, S. (2020). Detection of transmission change points during unlock-3 and unlock-4 measures controlling COVID-19 in India. *Medrxiv.*

Monaghan, C., Bizumic, B., Williams, T., & Sellbom, M. (2020). Two-dimensional Machiavellianism: Conceptualization, theory, and measurement of the views and tactics dimensions. *Psychological Assessment,* 32(3), 277–293.

Pandey, D., Aswari, A., Taufiqurakman, M., Khalim, A., & Azahrah, F. F. (2021). System of Education Changes Due to COVID-19 Pandemic. *Asian Journal of Advances in Research,* 10-15.

Pandey, D., Babu, S. Z. D., Noibi, T. O., Pandey, B. K., Dey, P. T. D., Kassaw, C., Kadam S. A., Canete, J. J. O. (2020). A study on novel covid-19 epidemic threat for universal physical situation in 2020. *Asian Journal of Advances in Research,* 27-33.

Pandey, D., Islam, T., Magray, J. A., Gulzar, A., & Zargar, S. A. (2021). Use of statistical analysis to monitor novel coronavirus-19 cases in Jammu and Kashmir, India. *European Journal of Biological Research,* 11(3), 274-282.

Pandey, D., Islam, T., Malik, M. A. (2021). Novel coronavirus disease (SARS-COV-2): an overview. *Asian Journal of Advances in Medical Science,* 8-12.

Pandey, D., Ogunmola, G. A., Enbeyle, W., Abdullahi, M., Pandey, B. K., Pramanik, S. (2021). COVID-19: A framework for effective delivering of online classes during lockdown. *Human Arenas,* 1-15.

Paul, A., Chatterjee, S., & Bairagi, N. (2020). COVID-19 transmission dynamics during the unlock phase and significance of testing. *medRxiv.*

Rehman, U., Shahnawaz, M. G., Khan, N. H., Kharshiing, K. D., Khursheed, M., Gupta, K., ... & Uniyal, R. (2020). Depression, anxiety and stress among Indians in times of COVID-19 lockdown. *Community mental health journal,* 1-7.

Roy, A., Singh, A. K., Mishra, S., Chinnadurai, A., Mitra, A., & Bakshi, O. (2020). Mental health implications of COVID-19 pandemic and its response in India. The International journal of social psychiatry, 20764020950769. *Advance online publication.* https://doi.org/10.1177/0020764020950769

Saha, J., & Chouhan, P. (2021). Lockdown and unlock for the COVID-19 pandemic and associated residential mobility in India. *International Journal of Infectious Diseases, 104,* 382-389

Serafini, G., Parmigiani, B., Amerio, A., Aguglia, A., Sher, L., & Amore, M. (2020). *The psychological impact of COVID-19 on the mental health in the general population.*

Sharma, S., & Sharma, A. (2020, November). Twitter Sentiment Analysis During Unlock Period of COVID-19. In *2020 Sixth International Conference on Parallel, Distributed and Grid Computing (PDGC)* (pp. 221-224). IEEE.

Shi, L., Lu, Z. A., Que, J. Y., Huang, X. L., Liu, L., Ran, M. S., ... & Shi, J. (2020). Prevalence of and risk factors associated with mental health symptoms among the general population in China during the coronavirus disease 2019 pandemic. *JAMA network open, 3*(7), e2014053-e2014053.

Shuler, R. L., Koukouvitis, T., & Suematsu, D. (2020). Partial unlock for COVID-19-like epidemics can save 1-3 million lives worldwide. *medRxiv.*

Stedman, M., Davies, M., Lunt, M., Verma, A., Anderson, S. G., & Heald, A. H. (2020). A phased approach to unlocking during the COVID-19 pandemic—lessons from trend analysis. *International journal of clinical practice, 74*(8), e13528.

Sundin, E. C., & Horowitz, M. J. (2002). Impact of Event Scale: psychometric properties. *The British Journal of Psychiatry,* 180(3), 205-209.

Thakur, V., & Jain, A. (2020). COVID 2019-suicides: A global psychological pandemic. *Brain, behavior, and immunity.*

Van Bortel, T., Basnayake, A., Wurie, F., Jambai, M., Koroma, A. S., Muana, A. T., ... & Nellums, L. B. (2016). Psychosocial effects of an Ebola outbreak at individual, community and international levels. *Bulletin of the World Health Organization, 94*(3), 210.

Walker, P. G., Whittaker, C., Watson, O. J., Baguelin, M., Winskill, P., Hamlet, A., ... & Ghani, A. C. (2020). The impact of COVID-19 and strategies for mitigation and suppression in low-and middle-income countries. *Science, 369*(6502), 413-422.

Wang, C., Pan, R., Wan, X., Tan, Y., Xu, L., Ho, C. S., & Ho, R. C. (2020). Immediate psychological responses and associated factors during the initial stage of the 2019 coronavirus disease (COVID-19) epidemic among the general population in China. *International journal of environmental research and public health, 17*(5), 1729.

Wang, C., Pan, R., Wan, X., Tan, Y., Xu, L., McIntyre, R. S., ... & Ho, C. (2020). A longitudinal study on the mental health of general population during the COVID-19 epidemic in China. *Brain, behavior, and immunity.*

Xiang, Y. T., Yang, Y., Li, W., Zhang, L., Zhang, Q., Cheung, T., & Ng, C. H. (2020). Timely mental health care for the 2019 novel coronavirus outbreak is urgently needed. *The Lancet Psychiatry, 7*(3), 228-229.

Xiong, J., Lipsitz, O., Nasri, F., Lui, L., Gill, H., Phan, L., Chen-Li, D., Iacobucci, M., Ho, R., Majeed, A., & McIntyre, R. S. (2020). Impact of COVID-19 pandemic on mental health in the general population: A systematic review. *Journal of affective disorders,* 277, 55–64. Advance online publication. https://doi.org/10.1016/j.jad.2020.08.001.

About the Editors

Dr. (h.c) Digvijay Pandey
Acting Head of Department in the Department of Technical Education
Kanpur, Government of Uttar Pradesh, India
Email: digit11011989@gmail.com

Dr. (h.c) Digvijay Pandey is currently working in the Department of Technical Education in Kanpur, India as an acting HOD in the IT Department. He obtained his M Tech with honours in digital systems and design at the KNIT Sultanpur (Dr. A.P.J. Abdul Kalam Technical University, Uttar Pradesh and Lucknow), India, in 2016. He obtained his first degree in BTech with honours at the IERT Allahabad (Dr. A.P.J. Abdul Kalam Technical University, Uttar Pradesh and Lucknow), India, in 2011. In 2012, he joined TCS (IT analyst) and worked on various US/UK/Canada projects till 2016 and also worked at IERT Allahabad as a faculty member. He has more than eleven years of experience in the fields of industry and teaching.

He has been on the review board of many reputed journals like Springer, Wiley, Sci, IEEE, and Scopus indexed journals. He has more than 90 publications in reputed journals, including Springer, Inderscience, and Science Direct (Elisever) (SCI and Scopus Indexed Journals), more than 15 book chapters, and 6 patents. He has received numerous awards, including the PM Scholarship, RajyaPrushakar, NationalPlayer, NCC C Certificate, NSS, and many others. He has been a session chair at various conferences such as IEEE, Springer, and the International "Conference on "Sustainable Business Management Practices and Social Innovation" conducted by SRDC – Vidhyanidhi, CMS, and Jain (Deemed-to-be University).

Prof. Binay Kumar Pandey

Assistant Professor in Department of Information Technology
of Govind Ballabh Pant University of Agriculture and Technology,
Pantnagar, Uttrakhand, India,
Email: binaydece@gmail.com

Prof. Binay Kumar Pandey is currently working as an Assistant Professor in the Department of Information Technology at Govind Ballabh Pant University of Agriculture and Technology in Pantnagar, Uttarakhand, India. He obtained his M. Tech with a specialisation in bioinformatics from the Maulana Azad National Institute of Technology, Bhopal, M.P., India, in 2008. He obtained his first degree in B. Tech at the IET Lucknow (Uttar Pradesh Technical University, Uttar Pradesh and Lucknow), India, in 2005. In 2010, he joined the Department of Information Technology at the College of Technology at Govind Ballabh Pant University of Agriculture and Technology in Pantnagar as an Assistant Professor and has worked on various UG and PG projects till date.

He has more than ten years of experience in the fields of teaching and research. He has more than 40 publications in reputed journals, including Springer, Inderscience, (SCI, Scopus Indexed Journals, and others) and 3 patents. He has many awards, such as the PM Scholarship, etc. He was session chair at the IEEE International Conference on Advent Trends in Multidisciplinary Research and Innovation (ICATMRI-2020) on December 30, 2020, organised by Pankaj Laddhad Institute of Technology and Management Studies in Buldhana, Maharashtra, India, and is a reviewer of SCI/Scopus indexed journals for Springer, Wiley, etc.

Dr. Nidhi Sindhwani (PhD)

Assistant Professor, ASETD, Amity University, Noida, India
Email: manchandanidhi804@gmail.com

Dr. Nidhi Sindhwani is currently working as an Assistant Professor at the Amity School of Engineering and Technology (Delhi), Amity University, Noida, India. She has teaching experience of more than 14 years. She received her B.E. degree in Electronics and Communication Engineering in 2004 and her M.E. degree in Signal Processing in 2008, both from Maharishi Dayanand University, Rohtak, India.

She has received her PhD (ECE) from UCOE, Punjabi University, Patiala, India in 2018. She is also a life member of the Indian Society for Technical Education (ISTE). She has published numerous papers with a good impact factor in reputed international journals and conferences. Her current research interests include Wireless Networks and Communication, IOT, Machine Learning, and Signal Processing.

Dr. Rohit Anand (PhD)

Assistant Professor, Department of Electronics
& Communication Engineering,
G.B. Pant Engineering College (Government of NCT of Delhi),
New Delhi, India
Email: roh_anand@rediffmail.com

Dr. Rohit Anand is currently working as an Assistant Professor in the Department of Electronics and Communication Engineering at G.B.Pant Engineering College (Government of NCT of Delhi), New Delhi, India. He has done his Ph.D. (ECE) from IKG Punjab Technical University, Kapurthala, Punjab, India. He has teaching experience of more than 19 years, including UG and PG courses. He is a life member of the Indian Society for Technical Education (ISTE). He has published 6 book chapters in reputed books, 12 papers in Scopus/SCI Indexed Journals and 1 patent.

He has presented various research papers at national and international conferences. He is presently a reviewer for many of the reputed and highly indexed international journals. He has chaired sessions at two international conferences. His research areas include IoT, Wireless Communication, Electromagnetic Field Theory, Antenna Theory and Design, Image Processing, Optimization, Optical Fiber Communication, etc.

Prof. (Dr.) Vinay Kumar Nassa (PhD)

Professor (AIDS), Department at Sharad Institute of Technology College
of Engineering, Ichalkaranji, Kolhapur, Maharashtra, India
Email: vinay_nassa@rediffmail.com

Prof. (Dr.) Vinay Kumar Nassa is currently working as a Professor (AIDS) at the Sharad Institute of Technology College of Engineering,

Ichalkaranji (Kolhapur-Maharashtra-416121). He received his B.E (Electronics) degree from Nagpur University. He has two postgraduate degrees (M.C.A. from M.D.U. Rohtak in 2004 and M.Tech (ECE) from P.T.U. Jalandar in 2007). He also earned a PhD in computer engineering and electrical engineering. He has most of his experience in teaching M.C.A. & M. Tech. In addition to this, he is a member of various professional bodies like CSI, IEEE, ISTE, & IANGE etc.

He is on the panel of reviewers of many national and international journals. He has 26 years of teaching and 5 years of industrial experience. He had experience of establishing new institutes, NBA & NAAC accreditation, and worked as Director/Principal of various institutions. His major research areas are AI, Machine Learning, Mobile Adhoc Networks, and Deep Learning India.

Dr. Pankaj Dadheech (PhD)

Associate Professor & Dy. HOD in the Department of Computer Science & Engineering (NBA Accredited), Swami Keshvanand Institute of Technology, Management & Gramothan (SKIT), Jaipur, Rajasthan, India.
Email: pankaj@skit.ac.in

Dr. Pankaj Dadheech completed his PhD degree in Computer Science & Engineering from Suresh Gyan Vihar University (Accredited by NAAC with 'A' Grade), Jaipur, Rajasthan, India. He completed his M.Tech. degree in Computer Science & Engineering from Rajasthan Technical University, Kota and he has completed his B.E. in Computer Science & Engineering from University of Rajasthan, Jaipur. He has more than 16 years of experience in teaching. He is currently working as an Associate Professor & Dy. HOD in the Department of Computer Science & Engineering (NBA Accredited), Swami Keshvanand Institute of Technology, Management & Gramothan (SKIT), Jaipur, Rajasthan, India. He has published 21 Patents at Intellectual Property India, Office of the Controller General of Patents, Design and Trade Marks, Department of Industrial Policy and Promotion, Ministry of Commerce and Industry, Government of India. He has published 4 Australian Patents at Commissioner of Patents, Intellectual Property Australia, Australian Government. He has published 1 German Patent. He has also Registered & Granted 2 Research Copyrights at Registrar of Copyrights, Copyright Office, Department for Promotion of Industry and Internal Trade, Ministry of Commerce and Industry, Government of India.

He has presented 53 papers in various National & International Conferences. He has 57 publications in various International & National Journals. He has published 5 Books & 11 Book Chapters. He is a member of many Professional Organizations like the IEEE Computer Society, CSI, ACM, IAENG & ISTE. He has been appointed as a PhD Research Supervisor in the Department of Computer Science & Engineering at SKIT, Jaipur (Recognized Research Centre of Rajasthan Technical University, Kota). He has also guided various M.Tech. Research Scholars. He has Chaired Technical Sessions in various International Conferences & Contributed as Resource Person in various FDP's, Workshops, STTPs, and Conferences. He is also acting as a Guest Editor of the various reputed Journal publishing houses, Conference Proceedings and Bentham Ambassador of Bentham Science Publisher. His area of interest includes High Performance Computing, Cloud Computing, Information Security, Big Data Analytics, Intellectual Property Right and Internet of Things.

Head of Technical Committee Advisor

Dr. A. Shaji George has served the education and ICT industries for over 25 years. He is a member of the board of trustees and professor at Crown University International and its cooperating advisory board of trustees, and a special assistant to Pro Chancellor at Chartered Int'l Da Vinci University Inc., Nigeria, West Africa. As a Deputy Vice President (Administration), West Coast University International, Nigeria, West Africa, I collaborate with the advisory board of trustees. Trainer and Counselor of the National Human Rights and Humanitarian Federation-NHRF, Kerala, India. A recognised technical expert in ICT Systems, Networks & Telecommunication. designed and deployed large scale ICT projects. Having published more than sixty research papers in national and international journals with impact factors, his research interests include Wireless Networking, Cloud Computing, Image Processing, Big Data, IoT, and industrial automation systems.

Technical Committee Advisor

1. Dr/Prof Manvinder Sharma
CGC Chandigarh
Email: manvinder.sharma@gmail.com

2. Wegayehu E.
Director. International Relations and Partnership Directorate Office
Department of Statistics
Mizan-Tepi University, Ethiopia
Email: wegayehu.info@gmail.com

Index